# The Great Years

# *Gold Mining in the Bridge River Valley*

# The Great Years

Lewis Green

Tricouni Press
Vancouver

Second printing August 2002



**Canadian Cataloguing in Publication Data**

Green, Lewis, 1925-
The great years

Includes bibliographical references and index.
ISBN 0-9697601-3-2

1. Gold mines and mining—British Columbia—Bridge River Valley—History. 2. Bridge River Valley (B.C.)—Gold discoveries—History. 3. Bridge River Valley (B.C.)—History. I. Title.
FC3822.4.G74 2000 971.1′31 C00-910772-X
F1089.B8G74 2000

Available from:
Tricouni Press Ltd.
3649 West 18th Avenue
Vancouver, BC, Canada V6S 1B3
Phone and fax: 604-224-1178
Email: books@tricounipress.com

and from:
Gordon Soules Book Publishers Ltd.
1359 Ambleside Lane
West Vancouver, BC, Canada V7T 2Y9
Phone: 604-922-6588
Fax: 604-688-5442
Email: books@gordonsoules.com

Designed by Glenn Woodsworth
Set in Minion and Syntax by Joy Woodsworth
Printed and bound in Canada on acid-free paper by Hignell Printing Ltd.

Front cover and title page photo: the day crew at Bralorne Mines, October 30, 1939. Photo by W.V. Ring.

*Dedicated to the memory of the pioneers, especially*
*David Sloan,*
*Arthur and Delina Noel,*
*and Ira Joralemon.*
*All were firm believers in the great potential*
*of the Bridge River gold camp.*

# Contents

Panoramic view looking east across Cadwallader Creek, which flows from right to left near the bottom of the photo. The Bendor Range forms the skyline above the valley. The Bralorne mine is in the lower left of the photo, below the road. The mine office buildings are just above and to the right, where the two conspicuous roads join. The light coloured patch high on the hillside above and slightly right of the office area is the Lorne mine. The Bralorne No. 1 townsite occupies much of the central part of the panorama above the main road. In the right part of the photo, between the photographer's writing and the main road is the Bralorne No. 2 townsite. Pioneer is just visible in the "V" at the head of Cadwallader Creek on the far right of the photo. Photo taken October 1939 by W.V. Ring.

# *Preface and Acknowledgments*

Gold mining was one of a few industries, legal or otherwise, that boomed throughout the Great Depression, its salvation an unlimited demand for a product selling at a fixed price. In contrast, British Columbia's base metal mines, victims of falling prices and poor demand, struggled to survive with the value of their 1932 production falling to less than a third the 1929 figure. Gold's lure drew the venturesome to remote mining camps, some in hopes of a job and others on the lookout for business opportunities or the chance to prospect for their own motherlode.

I never worked in the Bridge River's mines, but my interest in gold mining dates back to the 1950s and 1960s when, as a government geologist mapping in the Yukon, I visited and reported on numerous small placer mining operations active at that time. Many of the owners had come north during the Dirty Thirties, and after starting out as hand miners, had

gradually added mechanical equipment. Over the years a few operations had become highly profitable, but most provided a modest income at best. The miners, fiercely independent and beholden to no one, were deservedly proud what they had accomplished. I consider myself privileged to have known such remarkable people.

In 1967 I had my first glimpse of the Bridge River valley on a flight from Whitehorse to Vancouver aboard a small floatplane. South of Prince George, we followed the Fraser River to Dog Creek before leaving the river to fly directly towards Pemberton, only to find ourselves trapped beneath a descending cloud cover. Forced lower, we caught a glimpse of Carpenter Lake and were able to follow Cadwallader Creek past the mines and toward McGillivray Pass. By now, with the plane well below the valley walls, we could look *up* at a hunting party with packhorses on the ridge to our north and at a telephone line strung between poles along the crest to the south. In the pass itself we briefly plunged into cotton-wool clouds before breaking through into the clear above Anderson Lake and the rail line to Pemberton.

In 1972 I made my first visit on the ground. Bralorne, with the mine closed for a year, was already looking shabby. Pioneer, closed in 1960, was in ruins. It was difficult to appreciate that this remote valley, encircled by high mountains and lacking a direct road connection until 1955, had once been a prosperous mining camp, home to some 3,000 people. Its abandonment and rapid decay reminded me of what I had witnessed in the Klondike after the gold dredges were shut down in 1966.

My interest in the history of Bridge River's mines - once so important and now all but forgotten - was kindled by the late W.A. Hutchings, who opened the Bank of Montreal's Bralorne branch in the summer of 1934 and in 1937 returned to the Yukon as manager of the Dawson City branch. A man with a great sense of history, he helped me with an earlier book on gold dredging in the Klondike and, in the process, regaled me with recollections of his Bridge River days. Unfortunately, he passed away in 1977 before I began interviews for this book, but I have had the use of some of his writings and a photograph album. The latter includes both scenes from his years at Bralorne and a collection of older material given him by Mrs. Delina Noel.

The framework for my work is based on the annual *Minister of Mines* reports, company annual reports, technical publications and newspapers, particularly the *Bridge River-Lillooet News.* Using these, I have tried to weave in the recollections of individuals willing to be interviewed and to share their memories and photographs. Like their Yukon counterparts, many who came to the Bridge River in the Dirty Thirties were remarkable people who persevered, with many adventures along the way. Their story deserves to be recorded, and I hope my attempt to do this would meet with their approval.

Much of my work was done in the library of the University of British Columbia, and as with my earlier books, I consider myself a beneficiary of the policy that makes the stacks of this excellent facility available to the general public.

At the Legislative Library in Victoria I was permitted to examine and make notes from copies of the *Bridge River-Lillooet News* and other earlier Lillooet newspapers held in their stacks. These have since been microfilmed and can now be studied at the British Columbia Archives. Local newspapers are a wonderful source of historical information, and it is indeed fortunate that through the years this library has both collected and preserved so much of this material.

In the early 1980s Brian Young of the British Columbia Archives was most helpful in locating and arranging permission for me to examine corporate records of mining companies, both active and defunct, involved in the Bridge River story.

I have attempted to contact the copyright holders of unpublished and previously published material, and I apologize to those I may have overlooked.

Permission to quote excerpts from Emma de Hullu's *Bridge River Gold,* Ira B. Joralemon's *Adventure Beacons* and John Stanton's typescript prepared for *Never Say Die!* is gratefully acknowledged.

Photo credits are given with each photo, but special thanks go to C.E. Cleveland, E. de Hullu, W. StC. Dunn, D. Ingram, I. Howard, M. Jukich, G.B. Leech and B. Orgnacco for their help in locating them.

Many thanks to Glenn and Joy Woodsworth, owners of Tricouni Press, for their willingness to publish a manuscript that may be of limited interest to those without ties to the Bridge River.

Finally, there is the continual encouragement I have received from my family, especially my wife, during this project.

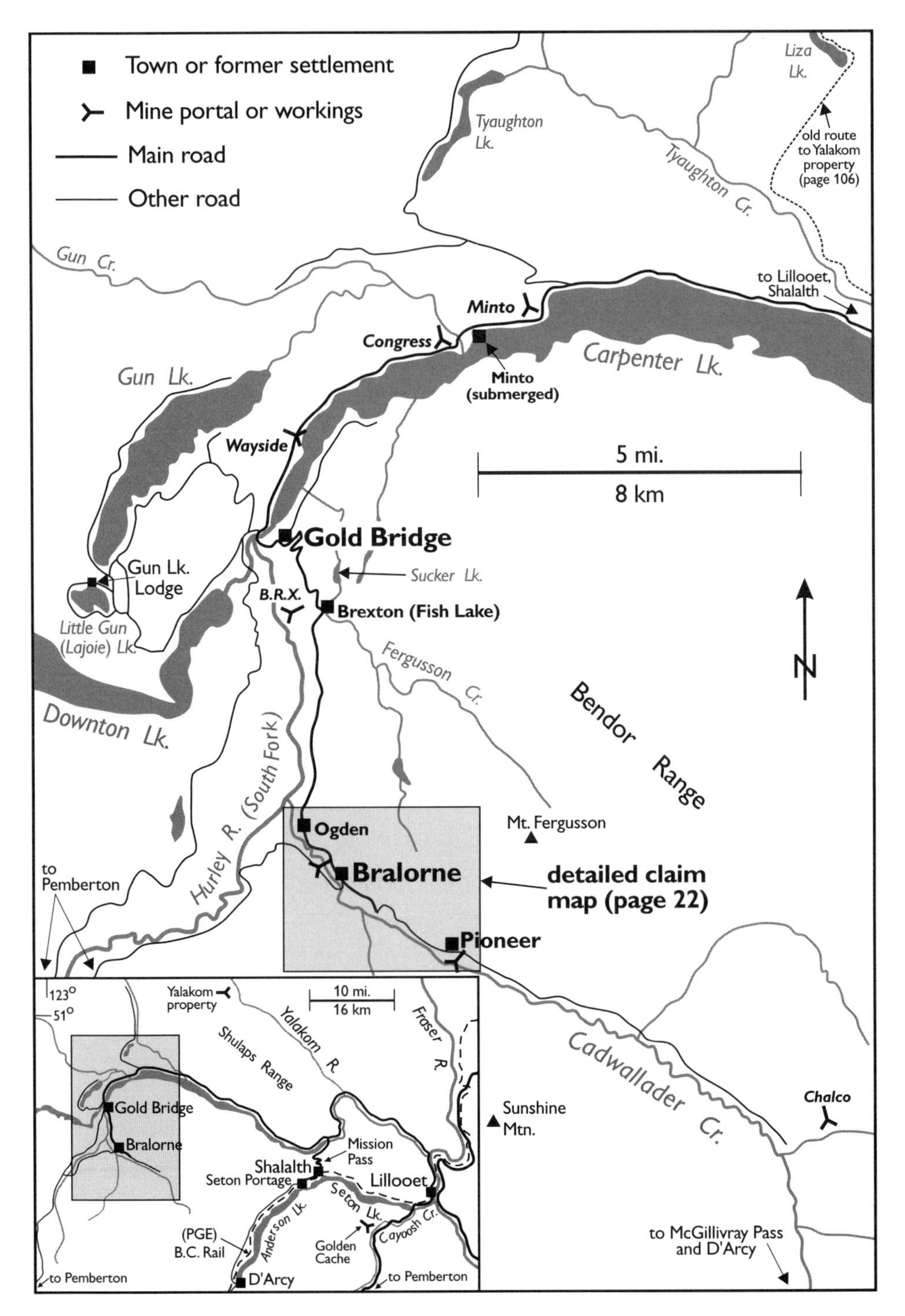


Bridge River valley. See p. 58 for location with respect to Vancouver and the main 1928 transportation network.

# *The Setting*

Fall 1933 found Al Abbott, a prairie boy in his early twenties, hurrying to British Columbia's Bridge River district where his brother had found a job for him. Until now the Great Depression had brought nothing but hard times, but finally there was a prospect of regular wages and decent food, and if it meant going to a remote mining camp to find them, so be it.

Two years earlier, Abbott and his younger brother had left their home in southern Saskatchewan and drifted west, working at the harvest or whatever jobs they could find. They were soon separated, losing touch with one another, although Al, not letting on but fooling no one, kept writing their mother that things were going well for them both. The end of the 1932 harvest found him near Three Hills, Alberta, low on cash and with no prospect of a job. He survived by sleeping in his car, a Model T Ford coach, and cutting back to two meals a day, toast and coffee in the morning and soup and crackers in the evening. It went on that way until after Christmas, in temperatures that sometimes dropped to 40°F below zero. Finally there was a job at an open-pit coal mine where the pay of $25 a month and good grub made up for the poorly heated bunkhouse where the wind whistled through cracks in the walls and blankets were often white with frost by morning. It was farm work again in the spring and

The main street of Lillooet, about 1934. Photo by W.A. Hutchings.

summer of 1933, but now, the masquerade over and contact with his brother restored, things were finally looking up.

Abbott set out for the Bridge River on the Crowsnest route through southern British Columbia. Misfortune tagged along, beginning when the brakes on his Model T failed on the steep descent from the summit east of Grand Forks and Abbott, terrified, just managed to keep the car on the road in a wild ride to the bottom. Short of money, he found a friendly garage man who agreed to fix the brakes in return for a few days work around the place. Then, on reaching Lillooet, a dusty, weather-beaten town sprawled on a bench above the Fraser River, he learned there was no direct road connection to the Bridge River and that it would take the little cash he had left to ship his car by rail across the 15-mile gap to Shalalth on the shore of Seton Lake. He paid, saw his car loaded onto a flatcar and boarded the gas-car, a self-propelled baggage and passenger unit that the Pacific Great Eastern Railway used on the shuttle.

The gas-car rattled along a few feet above Seton Lake, lurching one way and the other on sharp curves as the rails, laid on a narrow bench blasted from bedrock or cut through active slides, followed the twisting shoreline. The reason for the road gap was obvious: building one would be a major undertaking certain to increase the risk from both mud and snowslides on the precipitous slopes rising abruptly from the lake.

Pacific Great Eastern Railway gas-car 107 at Lillooet, about 1950. Photo courtesy of K. Anderson.

Shalalth is on a narrow strip of usable land by the lake. The settlement held an Indian reserve, some farms and, a little beyond, a rail siding, freight sheds and garages of the trucking companies. The Depression was here too, marked by a silent group of buildings that were part of a major hydroelectric project intended to supply the Vancouver area. The B.C. Electric Company's plans had called for water from the Bridge River to be diverted through a 2½-mile tunnel for delivery at a head of almost 1,000 feet to turbines and generators installed along the shore of Seton Lake. Work begun in 1926 had been shut down in 1932 after completion of the tunnel, and the company had later announced work would not resume until the demand for electricity picked up.

From the railhead at Shalalth, six miles of switchback road climbed 3,000 feet to reach the summit of Mission Pass, the entrance to the Bridge River valley. Abbott, impatient to start his new job, decided to risk starting up the road despite being low on gas and, flat broke, unable to do anything about it. The climb was on a dry, south-facing slope mostly open and grass-covered, but with a few scattered clumps of good-sized ponderosa pine. As he drove, Abbott was too worried about running out of gas to appreciate the spectacular views of Seton and Anderson lakes unfolding below him. He even backed up some of the switchbacks in the hope that the Model T's gravity fuel system would deliver the last drops in the tank, but it was no use and the engine sputtered and died. With no room to park on the narrow road he managed to get his car turned around, coasted back down to Shalalth and was asleep in it when his new boss, Red Doffner, came by and picked him up. At the time, Doffner and his mother had a small grocery store at Ogden near the Bralorne mine and a truck that Abbott would be driving on the Shalalth run.

Riding with Doffner, Abbott realized that he had been within a single switchback of the Mission Pass summit. Beyond, the road descended a wooded slope nearly 2,000 feet to reach the Bridge River valley about two miles upstream from the point where the river tumbled into an impassable canyon. The road ran along the south side of the valley for a short distance, crossed on a narrow bridge and then ran for close to 30 miles along the north side of flats, in places close to a mile wide, formed by the meandering Bridge River. Along the way they passed a number of ranches and two good-sized mining properties, Minto and Wayside, both operating with crews of about 30 men.

The road recrossed the Bridge River just below the mouth of a large tributary, the Hurley River or South Fork, to begin the long climb up Cadwallader Creek to Bralorne and Pioneer mines. The South Fork Lodge lay just across the bridge and a short distance past it were the newly opened Goldbridge Hotel and the beginnings of a settlement. The road, rougher now, ran through timbered country and here and there Abbott caught glimpses of whitewater in a foaming canyon far below and of the towering snow-capped peaks ringing the valley. As they went along Doffner pointed out side roads leading to other properties where crews were working. At last, 50 miles out, they reached Ogden, a store and a few rough buildings perched high up on a steep hillside – a next-to-impossible site, its sole advantage being proximity to the Bralorne property and potential customers.

Abbott soon had his first look at Bralorne and Pioneer, the two producing mines responsible for the boom. Both were bustling, the new construction underway suggesting both expansion and a long-term future. Everything was being done in a hurry, with truckers delivering equipment and supplies round the clock. At Bralorne, half a mile beyond Ogden, the crew numbered close to 130 men. Camp buildings were on a bench about 150 feet above Cadwallader Creek, the main portal of the mine part way down the slope, and the mill and power plant on a flat just above creek level. Mill capacity had recently been doubled to 200 tons per day and ore trains, powered by battery locomotives, shuttled in and out of the mine. Some of the buildings were close to five years old; new ones being rushed to completion included a community hall, an office building and a combined doctor's residence and emergency hospital. These, together with al-

Bralorne, buildings on the bench above the mill, about 1934. Photo by W.A. Hutchings.

Pioneer Mine, 1933. Photo by Leonard Frank, Public Archives of B.C. A-06588.

terations and additions to both mill and power plant, gave the camp a raw, unfinished look.

At Pioneer, another 2½ miles upstream, 245 men were at work, the mill was treating over 300 tons per day and much new ore* had been discovered. Unlike Bralorne, there were no good-sized benches and most of the buildings were crowded into the now narrow, V-shaped valley of Cadwallader Creek. New additions included a bunkhouse, recreation hall, school, directors' guest house and four small houses.

Seeing was believing: this boom was not about to fade overnight. The desperation Abbott had known on the prairies lifted, but the Fates still had another trick to play on him. On his first run to Shalalth about a week later, plans to pick up his Model T were abandoned when he found the locals had beaten him to it and stripped everything of value. He would stay in the Bridge River area for two years, working at a variety of jobs, the last

* term described in Glossary, p. 242.

as a mucker at Pioneer mine. On leaving, he went back to working at whatever jobs were available but it was easier now; his time in the Bridge River had both seen him through the worst of the Depression and given him the confidence needed to tackle anything.

Abbott had arrived in the Bridge River just as things were moving into high gear with monthly production of the two mines close to 10,000 ounces of gold worth about $30 US an ounce. As of January 31, 1934, that price would rise to $35 US, worth more in real terms than $400 US gold in the devalued dollar of the 1990s. Most of the gold was shipped as bricks of varying weights, commonly in the 60-pound range, and photos of a stockbroker or other visitor hefting a gold brick made great publicity. Pioneer, prospering, had been paying dividends for three years, and Bralorne, with a new-found bonanza, would soon be joining it. Beyond the mines, a strip of favourable ground some 20 miles long had been blanketed by mineral claims* and new companies, flush with money raised in Vancouver, were impatient to explore their holdings. For the moment no one seemed concerned that the new companies were not producing anything; gold showings were turning up everywhere and given time some of these could well be winners too. Reality would come later and the "might have beens" are still there today. None of them have been bonanzas, although two, Minto and Wayside, have produced modest amounts of gold.

The boom had been a long time coming. Placer* gold had been discovered in the Bridge River drainage in 1858, and most of the important surface showings on what would in time become the Pioneer and Bralorne properties had been discovered and staked before the turn of the century. Attempts to develop mines began then and continued through the years, but all had failed and the Bridge River was dubbed "the valley of dead mines"[1] until one man, David Sloan, had the vision and the persistence to turn the Pioneer property into a successful mine. Others, following Sloan's lead, gambled on the Bralorne ground and, within a few years, it became the larger of the two mines.

The Bridge River's gold was won from quartz* veins,* remarkable structures seldom more than six feet wide but so continuous that they could be followed for thousands of feet, both horizontally and vertically, in the labyrinth of underground workings. The veins, filling fractures in dense, hard rock, had been in place for close to 90 million years,[2] long before man appeared on Planet Earth let alone coveted the soft, lustrous metal. Over time, surface outcroppings of the veins, attacked by water and the oxygen of the atmosphere, had crumbled to release the gold now found in the placer deposits.

For thousands like Al Abbott who flocked to the Bridge River in the bitter years of the Great Depression, it was a never-to-be-forgotten adventure. Some were down to their last few pennies when they reached their goal. A few luckier ones had jobs waiting for them but most would have to survive as best they could for weeks or even months until something turned up. Some were defeated, often through no failing of their own, and had to leave worse off than when they arrived, but many others managed to hang on, endure the hardships and succeed. Each individual's story is different, but the same thread runs through many of them: the trip to the Bridge River, the struggle to find work and, for some, the lucky breaks. With it all was something nobler – the willingness of people with next to nothing to share their food or the shelter of a scrap of canvas with someone even less fortunate. Little things, done to help out, made a lasting impression on many of the recipients who, even today, marvel at the generosity shown them by total strangers.

The boom ended in 1942. Many of the younger men joined the armed forces, and others, too old or medically unfit, drifted away to take better paying jobs in the wartime industries springing up in the Vancouver area and elsewhere. Mining continued at Pioneer until 1960 and at Bralorne until 1971. The latter's closing ended 47 years of continuous gold mining in the Bridge River. During that time the industry had produced over four million ounces of gold, provided jobs for a great many and great wealth for a few, and served as a training ground for engineers, geologists and miners active in the mineral industry of the province and, indeed, throughout the world. Intangible, but perhaps most important of all, it had kept alive the hope of better things to come during the darkest days of the Great Depression.

David Sloan, mine-maker and managing director of Pioneer (right), Will Haylmore, Deputy Mining Recorder, and Dr. A.R. Thomson, Pioneer director (left). Photo courtesy of D. Sloan.

# *Beginning Years*
## *1858–1932*

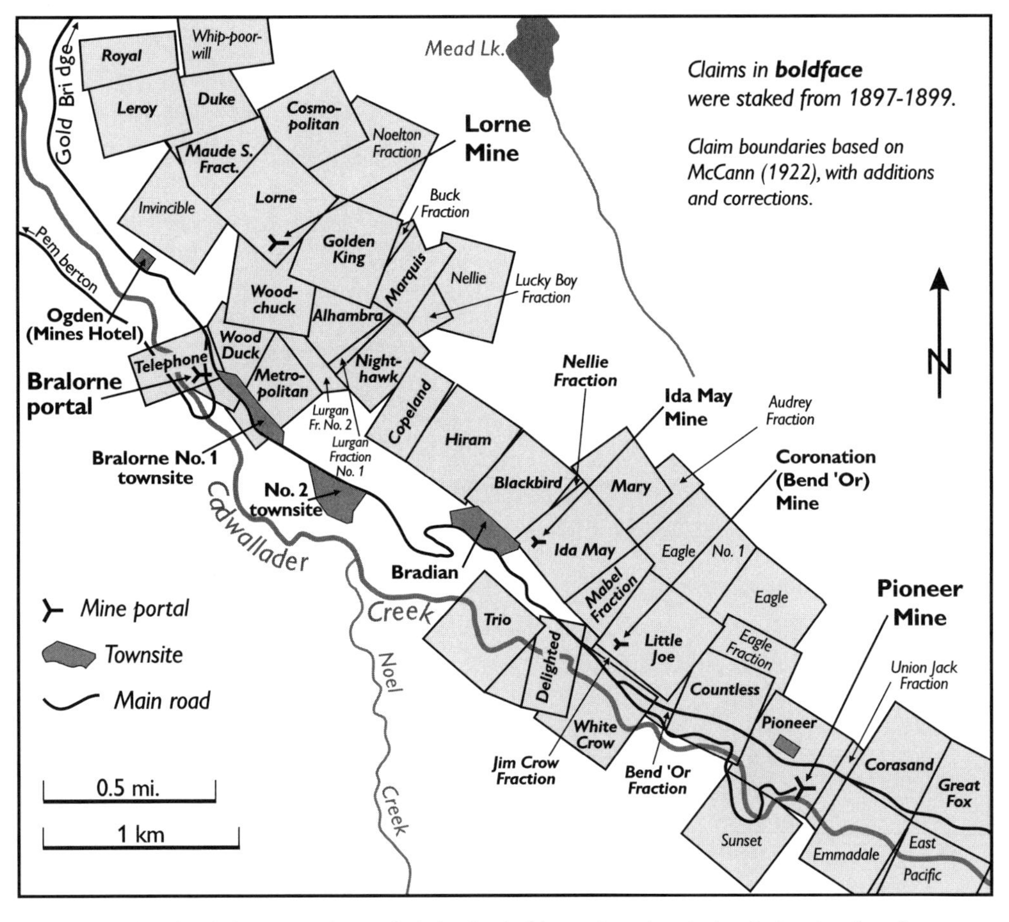


Sketch of the main claims (lightly shaded boxes) in the Cadwallader Creek valley. See p. 124 for a more regional view of the claim groups as they were in 1936.

# *Early Lode Mining, 1896–1923*

The Bridge River's first lode* or quartz claims were staked in 1896 after $25,000 paid for the Golden Cache property near Lillooet awakened prospectors to the potential rewards of hard rock* mining.

Events leading up to the Golden Cache discovery went back to 1886 when a new, totally unexpected placer find on Cayoosh Creek had provoked F. Soues, the government gold commissioner based in Clinton, into commenting:

> It seems almost incredible that this creek within an hour's walk of the town of Lillooet should have been passed by for a quarter of a century by hundreds, aye thousands, of the best practical white miners and prospectors on the Pacific Coast, and now at this late date the prize falls into the hands of the Chinese. Mr. Smith has purchased dust from it this year about $12,000, and estimates an equal amount in the miners' hands, in all say $24,000 for this season alone, and, of course, the creek is not worked out, and is one instance, at least, that may be placed on record that prospectors and miners in the past have not been by any means thorough in their search.[1]

The following summer two prospectors grubstaked to explore Cayoosh Creek for lode deposits found a large boulder of quartz and slate on a Chinese miner's placer claim and, on breaking it by heating in a fire and pouring water on it, disclosed visible gold. Prospecting the steep slope above, they discovered and staked an area where similar specimens occurred in abundance. A small amount of work was done on their Bonanza property, and over the next few years other lode claims were staked, only to be abandoned without any attempt at exploration.[2]

Arthur F. Noel, destined to change all that, first appears in the Lillooet mining records as staker of the Gold Slide claim on Cayoosh Creek, located February 28, 1894. By one account Noel, about 25 years old at the time, had been lured north from Montana by the sight of a Cayoosh Creek placer nugget with quartz adhering to it, suggesting a nearby bedrock source.[3]

Later, Noel worked for a company developing the Bonanza property and on April 12, 1896, he and Joe Copeland, the Bonanza blacksmith, staked the first claims of what would become the Golden Cache property. Once again, versions differ; one has Noel recognizing the potential value of samples of gold-bearing quartz that Copeland brought into a hunting camp, while the second has the pair spending a Sunday prospecting the creek.[4] The Golden Cache showing, across the valley and much higher than the Bonanza, consisted of a quartz lens or ledge,* about 450 feet long and 20 feet thick at the widest point, exposed in the face of a near-vertical bluff some 1,700 feet above creek level.

A specimen sent to Vancouver came to the attention of J.M. Mackinnon, a city businessman, who hurried to Lillooet and, on behalf of himself and partners, bought the claims for $25,000. Noel stayed on to manage the property and towards the end of the year had a crew of twelve men driving a tunnel on the quartz ledge. By 1897, the Golden Cache had a new superintendent after Noel resigned and left for California on holiday. Some $20,000 was spent building a stamp mill* beside Cayoosh Creek and a three-rail gravity tramway to bring the ore down to it.[5]

The provincial mineralogist visited the property in late September 1897 and was disturbed to find sampling indicated that the core of the quartz lens was barren and better gold values were concentrated within two to three feet of the contact* with enclosing sedimentary* rocks. Unable to visit the underground workings, all blocked by "ore" waiting to be trammed* to the mill, his doubts fill a final paragraph:

> This property attained a very sudden and far-reaching notoriety, by reason of the fine specimens found, and it is sincerely hoped that this large body of quartz will yet be found to carry high enough values to permit remunerative returns.[6]

It never would, but the 1896 staking rush saw 220 claims recorded in the Lillooet district, up from a mere 27 the year before. Included were three claims in the Bridge River valley staked June 14, 1896, to cover a quartz vein exposed on the east bank of the Hurley River some 1½ miles upstream from its mouth. The stakers, John Marshall, Paul Santini and John R. Williams, recorded their claims, the Forty Thieves, Elephant and Ural, respectively, in Lillooet on July 2. The quartz vein, about three feet wide and standing out as a white slash across the face of an inaccessible cliff, must surely have been observed and commented on years before. It seems likely that the trio, knowing this and very much aware of the

$25,000 paid for the Golden Cache property, had hurried in to be the first to stake. The presence of gold could be confirmed by crushing and panning* partially decomposed vein material while bright blue specks of copper carbonate signalled the presence of silver-bearing copper minerals. Later the property would prove to be a will-o'-the-wisp with a few erratic high gold values and only traces of silver, but by then it had served to attract others to the area.

In 1897 prospectors flocked to the Bridge River, abandoning Cayoosh Creek for the time being. A total of 175 claims were recorded on Cadwallader Creek alone, with an additional 245 in the general area. For the entire Lillooet district the grand total was 1,135 claims, a five-fold increase from 1896. Even Caspar Phair, the government agent and mining recorder in Lillooet, got involved with his Berta claim, located July 3, 1897, adjoining the Forty Thieves group. Gold Commissioner Soues was appalled by it all:

> This year may be characterised as one of fevered and profitless excitement, with an indiscriminate staking and recording of alleged mineral claims, in total disregard of the sanctity of an oath and the requirements of the Mineral Act, working incalculable harm to the mining future of the District.[7]

No doubt Soues had a point and many of the claims were simply "moose pasture*" without any "mineral in place" as required by the Mineral Act, but on others preliminary trenching had already disclosed gold-bearing quartz veins. Indeed, what would later become the key claims of the Pioneer and Bralorne properties had already been staked. The most work had been done on the Ida May group of three claims on Cadwallader Creek, where a vein had been stripped on the surface for 40 feet and followed underground for an additional 50 feet in a tunnel or drift* driven along it. The vein was three feet wide and samples taken across it averaged $40 per ton, or not quite two ounces of gold per ton at the long-established price of $20.67 an ounce. Ore of that grade, if there was enough of it, could surely be mined at a profit. Already there was the suggestion that a belt of gold-bearing veins, 10–12 miles long, might run from upper Cadwallader Creek to the mouth of the Hurley River.

The prospectors who staked in the summer of 1897 were a mixed lot, some acting on their own and others grubstaked to go in and prospect by backers who would hold an interest in any claims staked. One thing they had in common was the ability to travel and survive in a remote area, the difficult trip over Mission Pass serving to weed out the incompetents.

Under the regulations in force at the time, an individual holding a Free Miner's Certificate was allowed to stake a single claim on a vein or lode although, on making the required arrangements, he could also act as an agent for others. Claims, square or rectangular blocks with a maximum dimension of 1,500 feet on any side, were marked by two posts erected at the ends of a blazed location line. After staking, the locator was allowed 15 days, plus an additional day for each ten miles beyond the first ten, to travel to the recording office. Provided all regulations had been complied with, any earlier claim took precedence, a subsequent claim getting only any open ground within the area staked. Once recorded, a claim was in good standing for a year. Then it could be held from year to year by doing assessment work on the ground to the value of $100 and filing an affidavit describing it with the mining recorder. Beyond that, the holder of a claim, on fulfilling certain requirements, could apply for a Crown grant* which, in effect, turned the mineral and some surface rights into a piece of property recorded in the Land Registry.

The claim owners on Cadwallader Creek set to work prospecting for showings on their ground. Most had little or no money but a great deal could be accomplished with pick, shovel and sweat. Exposing a promising showing meant that the claim could be sold for a premium price, either in cash or perhaps in a combination of cash and stock in a company formed to develop it. There were few rock outcrops, most of the ground on the southwest-facing slope was either open grassland or covered by clumps of large Douglas fir or ponderosa pine. In addition, prospecting was complicated by local accumulations of volcanic ash or, elsewhere, by a veneer of glacial deposits as much as 20 feet thick. The gold-bearing quartz veins, highly banded and containing about 5 percent sulphide* minerals, tended to oxidize and crumble under surface conditions, and indicators in the form of rust-stained white quartz could sometimes be found on the open slopes or clinging to the roots of overturned trees. Favourable areas, suggested by either gold-bearing float* or the projection of a vein from an exposure nearby, were prospected by hand trenching using pick and shovel or, less commonly, by ground sluicing* when a stream of water powerful enough to strip away the overburden could be diverted across them.

By the end of the 1898 season, veins had been discovered on many of the important claims between the Lorne and Pioneer. Several groups of claims had been consolidated and one property, the Lorne, was in production in a small way. On another property, Bend 'Or Mines Limited planned to have a mill in operation in 1899. Gold Commissioner Soues, kept at his desk in Clinton either by paper work or a disinclination to rough it on a

The Lorne arrastra in 1912. Photo by A.M. Bateman, Geological Survey of Canada, No. 22329.

trip to the Bridge River, was left in the dark about it all, unable to find out what was going on despite repeated requests for information.[8]

At the Lorne, the owners installed an arrastra,* a primitive device made of local materials but handicapped by a very limited capacity. Introduced into the New World by the Spanish conquistadors, it consisted of a circular pit paved with carefully fitted stones and enclosed by a watertight wooden retaining wall with a vertical spindle or shaft mounted in the centre and horizontal arms extending from it. Drag-stones, connected to the arms by chains, crushed ore fed onto the pavement as the shaft rotated at a few revolutions per minute turned either by water power, as at the Lorne, or by animal power. Crushing was done in water and, by adding mercury, gold was recovered as an amalgam* of the two metals.

John R. Williams of Lillooet, staker of the Ural claim in 1896 and the Marquis in 1897, was responsible for the design and construction of the Lorne's arrastra. He was well qualified. Born in Ontario about 1843, he had built and operated water-powered flour mills both there and later in the

Cariboo and, on his way west, had observed and perhaps worked on arrastras in California. Six men worked on the Lorne project. The largest item was a ditch two miles long carrying water to the site from small lakes higher on the hillside. Initially, Williams planned to use a turbine wheel for power but when this proved unsatisfactory it was replaced by an overshot water wheel, 22 feet in diameter, with leather belting used to transmit power to a single arrastra, 14 feet across. Started up in early October 1898, the arrastra was run for just over a month with shutdowns for cleanups* every week to ten days. In all, an estimated 30 tons of ore was treated and gold worth about $1,000 recovered.[9]

In 1900, after a second arrastra powered by the same water wheel had been added, an article in *The Prospector* of Lillooet described the cleanup:

> The Lorne is a scene of activity and the arrastras swing around and around never stopping until cleanup day arrives, then the drags are lifted out of the tubs and the slime is run into sluice boxes, thence into slum pits where it is saved and will eventually be worked as concentrates.* When all the water is out of the tub the bed of the tub is carefully scraped and eventually cleaned off with a towel until it looks like a floor of polished marble. Then the crevices are started open and all the amalgam gold is taken out with spoons and scrapers and placed into pans and handed to Dan Hurley who finally pans and washes it clean.... In about three hours everything is ready again, the bed is carefully laid, the drags replaced, with chains attached to the arms, the belt is put on and round they go grinding out a fortune for the owners.
>
> There is gold all around you and capital is still reticent in investing its money into properties that are dividend payers from the grass roots. There is no doubt whatever of this district will be one of the biggest gold producers in the province.[10]

The Bend 'Or Mining Company, a new one in the stable of J.M. Mackinnon and holding three claims and fractions about two miles upstream from the Lorne, took a different, more ambitious approach to mining their gold. Following development work on their Little Joe claim in the summer of 1898, the company announced in October that a contract had been let to supply a stamp mill capable of treating 30 tons of ore per day. With the Golden Cache Mine on the point of collapse, Mackinnon was out to turn the Bend 'Or into his big mine.

Arthur Noel, back with Mackinnon, was in charge of work at the Bend 'Or. A comfortable camp had been set up and the crew of about a dozen men, working in two shifts, were driving an adit* on the Little Joe vein. The camp boasted an assay office that, when results were needed in a hurry, could probably run a sample in less than 24 hours. Noel, in Lillooet in mid-November, reported that the vein had been followed underground for more than 350 feet and was still going "as well defined and as rich as ever."[11]

The contract for the 10-stamp mill had gone to the Wm. Hamilton Manufacturing Company of Peterborough, Ontario. Some of the machinery needed to be specially designed to keep the weight of the heaviest piece to less than 350 pounds, considered to be the maximum that horses could move over the rough trail. The machinery alone would weigh 25 tons, and there would be other equipment and supplies needed for the mining operation. Noel held the contract to freight the outfit from Lillooet to the mine,

Steam Launch *Minnehaha* on Seton Lake, no date, but sometime between 1899 and 1903. Public Archives of B.C. D-03147.

apparently satisfied that, unlike the Golden Cache, there would be no shortage of ore for this mill.

Everything was done on a tight schedule. By February 9, 1899, the first 1,800 pounds of machinery was at the foot of Seton Lake where Fred Appleyard had a new steam launch, the *Minnehaha,* under construction. The craft, 40 feet long and with a nine-foot beam, had been built with a steam engine installed. Within an hour of launching, a boiler had been slapped in place, fired up and the boat taken for a trial spin on the lake. There was to be no time wasted; the first load of machinery was scheduled to leave the next morning and the rest was strung out along the road from the railhead at Ashcroft.

At the Mission (now Shalalth) 12 miles up Seton Lake, Arthur Noel had a crew of close to 50 men and about the same number of horses standing by, ready to start moving the loads over Mission Mountain just as quickly as the *Minnehaha* could ferry them. On the first section, a 3,000-foot climb to Mission Pass, the switchback trail was unsuitable for sleighs and most of the loads were carried on the horses' backs. At the pass, loads were transferred to sleighs for the 2,000-foot descent into the Bridge River valley where new sleighs and toboggans had to be built to replace the originals, now worn out. Plans to move the loads over the ice on the Bridge River had to be abandoned near the Big Horn Ranch, less than a third of the way to the intended landing at Sucker Creek, when the ice went out.

Freighting Bend 'Or supplies on Bridge River ice, 1899. Photo from W.A. Hutchings album.

During the two-week wait for the river to be free of ice, five boats, 30 feet long and six in the beam, were built from whipsawn lumber. Each could carry 3,000 pounds of cargo and, with four Indians poling, the upstream trip to Sucker Creek some 40 miles along the meandering channel took two days, and the return a mere half day.

By early May, everything had been landed at Sucker Creek but, by now, the snow had disappeared and the hardest part, ten miles over a miserable trail to the Bend 'Or, still lay ahead. Many of the loads were moved by packhorse but some of the larger pieces could only be moved on a stone-boat or go-devil. It was rough going and, at times, up to five horses were hitched together in single file to drag a load through a bad spot. Finally on June 12, 1899:

> The last load was drawn to the mill on a "go-devil" by Joe Russell, who had the horse decorated with evergreens, and on arrival at the mill three cheers were given by the men. Chas. Noel, Joe Russell and John Henry deserve credit for the way they worked in moving the machinery, and they kept things on the jump from the start.[12]

Work on the Bend 'Or property had gone on during the winter with the foundation for the mill completed, Cadwallader Creek dammed and a flume 400 feet long built to deliver water to the millsite at a head of 25 feet. If needed there would be ample power to operate a mill twice the size of the one planned. Towards the end of June 1899, Arthur Noel resigned as manager, although he still remained involved with the company. His place was taken by E. Jones, "lately of California . . . who is an experienced miner and mill man."[13]

There were delays when materials had to be freighted in to complete the tramway from the mine to the mill and, a few weeks later, some additional machinery for the mill. Finally, on the afternoon of August 24, the mill was started up. It run until September 15, treating 420 tons of ore, all of it drawn from a two-year old dump accumulated from 400 feet of drifting on the vein.

The Bend 'Or stamp mill, like an arrastra, recovered gold by amalgamation* with mercury. Ore, on delivery to the mill, was put through a jaw crusher and reduced to under two inches in size and together with water and mercury fed automatically into two five-stamp batteries. Here the stamps, weighing 850 pounds, were raised by a rotating cam and then dropped 7½ inches at the rate of 102 drops a minute to crush the ore

Bend 'Or mine, 1910. One of the portals can be seen on the left; the tramway carried the ore down to the mill near Cadwallader Creek. Photo by Wm. Fleet Robertson, B.C. Minister of Mines *Annual Report,* 1910.

against the face of an underlying mortar block. The ore, crushed until it would pass through a fine screen, was then run over amalgamating plates, gently sloping sheets of copper coated with mercury to which both free gold and amalgam would adhere. At intervals, cleanups were made by scraping the accumulated amalgam from the plates. In addition, tailings* from the plates were passed over crude tables covered with blankets, in an attempt made to save the sulphide minerals. The material recovered assayed about 9 ounces of gold per ton but, for the moment, there was no means of recovering it.[14]

A visitor to the Bridge River camp recorded his impressions in a letter published in Lillooet's *Prospector* of September 15, 1899:

> You can hear the Bend 'Or stamp mill from this place [the Lorne property], pounding away in great style and singing a song by itself, proclaiming to mankind it was made for Bridge river, and in Bridge river it will stay. A big cleanup is expected, and everyone will be overjoyed to hear of the company's success. When one goes over the trails and takes into consideration the distance this plant has been hauled ... he will come to the conclusion that somebody is entitled to credit for this. Go it Arthur! Your plucky.

J.M. Mackinnon, Arthur Noel and a representative of the mill manufacturer were on hand for the first cleanup. Carrying the gold with them they reached the landing on Seton Lake about noon on September 18:

> The steamer Minnehaha was on hand, decorated for the occasion, and conveyed the party and a few of our citizens who had gone up to meet them, to the foot of the lake, when the party drove to town. The rigs had been decorated for the occasion, by the drivers, and was much of a surprise to the party.[15]

After melting to drive off any remaining mercury, the resulting gold brick weighed 378.78 ounces valued at $16.23 an ounce for a total value of $6,147.59. The mill, started up again, ran until the night of December 1 when the service pipe from the flume line to the mill froze solid, leaving the operation without power. In total, the mill had operated for 79 days, treating 1,557 tons of ore and recovering 1,671.3 ounces of bullion,* the latter a mixture of gold, minor silver and traces of impurities.[16]

Later, in the company's report for the year ending January 31, 1900, the directors claimed that but for the early freeze-up the entire $30,000 cost of setting up the operation could have been recovered in the first season. As it was the vein had now been tested on two levels,* the new one driven 100 feet below the first, and, with the purchase of the adjoining Mabel fraction for $2,550 in cash, they considered they had about 8,000 tons of ore "which can be safely counted on running $20 to the ton."[17] With mining and milling costs estimated at just under $5.50 a ton, it appeared to offer a tidy profit. Despite this, there were financial problems. The original purchase agreement for the claims called for $50,000 in cash plus $300,000 in shares, and an attempt to sell 100,000 shares at their par value of 50¢ had raised just over a third of the required cash. Fortunately, the property owners had been talked into taking what cash there was plus and additional 66,100 shares in settlement.

Arthur Noel spent much of the last half of 1899 shuttling between Lillooet and the Bend 'Or, often with J.M. Mackinnon. Then in mid-December, Noel and his cousin Joe Russell set out for Vancouver, the latter planning to return to his home in Wisconsin for the winter while "Noel will be at the coast for a few days and will then visit Kamloops on more important business so rumour says."[18] Two weeks later *The Prospector* confirmed it; Arthur Noel had married Delina L'Itallienne of Lillooet in a Christmas Eve ceremony at the Roman Catholic Church in Kamloops:

> For some time it has been whispered that the above happy event was one of the future probabilities, but the good people of this town were pleasantly surprised on Christmas evening with the unexpected arrival of the happy couple.... After a pleasant journey, and the congratulations of Lillooet's population... the newly-wedded couple entertained their numerous friends at their future home... Mr. Noel is well and favorably known in the mining affairs of this section, and is at present closely identified with the Bend 'Or mines.... His bride is a charming young lady, eldest daughter of Mr. and Mrs. Joseph Letalien of this city, and who can claim the sunny hills of Lillooet as her native home.[19]

Delina, married at 19, had been born in Lillooet, the daughter of an Overlander who arrived in 1862. However, her mother had taken Delina and her sister to Quebec in 1894 and the family had not been reunited until their return to Lillooet in March 1898. Educated in a convent school, her background seemed an unlikely preparation for life in the wilds of the Bridge River. Arthur and Delina must have gotten to know one another between August 1898 and February 1899, when Noel's sister, Anna, was visiting from Merrill, Wisconsin. The two girls met when the stage bringing Anna to Lillooet stopped to pick up Delina at the L'Itallienne ranch at Fountain, a few miles out of town. During the fall and winter Arthur was in and out of town on trips to the Bend 'Or mine and the girls visited back and forth between Lillooet and the ranch. Then in early April 1899, the L'Itallienne family moved to town to take charge of the Excelsior House dining room.

For the moment the couple made their home in a house on Lillooet's main street and from there Arthur made many trips to the mine, the first in mid-January 1900. The mill, with a new manager, started up again in mid-March. Noel went in about once a month to be on hand for the cleanup and to bring the retorted* bullion back to Lillooet. In late July 1900, Noel took his bride and her sister, Louise, on a month-long trip to the Bridge River, travelling by way of Seton and Anderson lakes, up McGillivray Creek and through the pass with its miles of alpine meadows to Cadwallader Creek and the Bend 'Or. For Delina, it was a first look at the country destined to be her home for many years.

About 40 men worked at the mine that summer. By now the vein had been explored by drifts on three levels, the lowest roughly 200 feet below the surface showing. The width of the vein varied from about four feet at

Arthur and Delina Noel at the Bend 'Or mine, summer 1900. Photo from W.A. Hutchings album.

its best down to only a few inches, much too narrow to mine. In general, widths in excess of a foot could be mined if sampling indicated reasonably good gold values. Fortunately, much of the vein material was so fractured and altered that it could be mined by pick and shovel without the use of explosives. The company reported that in the 1900 mill run of 168 days some 3,243 tons of ore had been treated to yield 2,100 ounces of bullion containing 1,681 ounces of fine or pure gold. The cost of mining and milling was estimated at $5.13 a ton and the average assay of the tailings at an unsatisfactory $4.49 worth of unrecovered gold per ton.[20] It was discouraging; rather than increasing as anticipated, the grade of ore had fallen from the previous year when just over one ounce of bullion was recovered from each ton of ore treated.

Things went from bad to worse in the 1901 season. Started up in early March, the mine operated until mid-August, treating 1,200 tons of ore and recovering a mere $9,000 worth of bullion before everything was shut down and the company, out of working capital, went into liquidation.[21]

The outlook for lode mining in the Bridge River area was bleak. Arrastras,* with a low capital cost, were a poor man's game offering bare wages or, with luck, a bit more in return for a great deal of hard work. Mining on a larger scale seemed certain to fail without the investment of time and money to prove up the ore reserves needed to supply a larger, more efficient mill operated on a year-round basis. In the years that followed additional arrastras were built near the Lorne and on the Countless and Pioneer claims, and mills were built on the Lorne, Ida May, and Pioneer properties. By the early 1920s all had failed. Most of the mining claims, aside from Crown-granted ones, had been abandoned, leaving the ground open for staking, and only a handful of people still lived in the Bridge River valley.

During those discouraging years there were a number of developments, mainly in the form of government help, that, when the time was ripe, would make it easier to mine the Bridge River's gold. The first was a provincial government wagon road from Shalalth to Pioneer, built in 1912–13. In late October 1912, the first automobile, carrying seven people, made the trip over Mission Pass and along the Bridge River valley to Gun Creek. Then around the first of July 1913, the government road supervisor and his party could boast of breakfasting in Lillooet, having their car ferried to Shalalth aboard the *Britannia* and making the run to the Coronation, as the Bend 'Or was then known, in time for supper.[22] With

One of the early cars in the Bridge River brought in by J.Z. Lajoie, a colourful promoter with big dreams. Photo taken about 1912, from W.A. Hutchings album.

the new road, the cost of freighting supplies from Lillooet to the mines dropped from about $5 to $3 a hundredweight, a considerable improvement but still no bargain.[23]

The federal government made a contribution in 1912, too, appropriating $13,600 for a telephone system, consisting of a single iron wire strung on insulators nailed to trees, that, by late fall, had been completed from Lillooet to the Coronation.[24]

In 1915, the Pacific Great Eastern Railway (PGE), incorporated in 1912 to provide a Vancouver to Prince George rail link, opened the Squamish to Lillooet section of the line. Freight cars, loaded at the factory with mining or other equipment, could now be moved to Shalalth, using a rail barge across the Vancouver-to-Squamish gap. The company defaulted on its bonds in the winter of 1917–18, forcing the provincial government, as guarantor, to take over and operate the line, albeit with reluctance to spend money on what appeared to be a white elephant. Fortunately for its customers, PGE employees, expected to make do with a poor roadbed and poor equipment, showed great ingenuity in keeping the trains running and more or less on time.

In 1912, G.M. Downton, a land surveyor working in the Bridge River, had recognized and applied for the rights to another of the area's resources, its hydroelectric power potential. North of Shalalth, the Bridge River flowed at an elevation of close to 2,000 feet above sea level and, if its water were brought through a 2½-mile tunnel under Mission Mountain, a vast amount of power could be developed utilizing the 1,200-foot drop to the level of Seton Lake. Nothing came of the scheme until 1925 when the company formed by Downton to hold the rights was bought by the British Columbia Electric Company. In 1926, B.C. Electric began preliminary work on the project and in 1927 let the contract for the first tunnel. For the moment there was only indirect aid to mining in the form of much-needed improvements to the road over Mission Pass. Later, as the mines came into their own, power from the B.C. Electric plant would be available throughout the area.

The changes cut both the isolation and the cost of mining exploration in the Bridge River. If there was a profitable gold mine to be developed, something no one could be certain of, it would take triple synergy to pull it off: the right man, on the right property and at the right moment. Dave Sloan, the man who was to make it happen, took charge of the Pioneer mine in July 1924.

Frederick Kinder's one-stamp mill, the first mill at Pioneer.
Photo by W.A. Hutchings.

# *Pioneer: Dave Sloan Brings in a Mine, 1924–1932*

In 1931, 34 years after the Pioneer claim was staked on September 6, 1897, the property came into its own when Pioneer Gold Mines of B.C. Limited paid an initial dividend of 3¢ a share. In those earlier years there had been disappointment after disappointment, but after Dave Sloan took charge in July 1924, Pioneer had developed steadily, paying its way with gold bricks turned out at the mill. It was just the beginning; even better days lay ahead.

In the summer of 1898, W.F. Allen and Harry Atwood, certain their Pioneer vein was the equal of any of the new finds, let a contract for construction of an arrastra capable of handling two tons of ore per day. The arrastra started up on September 1, ran for several weeks but, with something wrong, only a small amount of gold was recovered. Atwood, certain he could do better, took over after the builders left and ran it for a few hours, recovering $20 worth of gold.[1] He tried again the following summer, but results must have been disappointing as, unlike the Lorne, the Lillooet newspaper makes no mention of cleanups.

In 1900 Allen and Atwood tried a new approach, striking a deal with Frederick H. Kinder to install a three-stamp prospecting mill on the claim. Kinder, an erstwhile house painter and since 1899 the operator of the steamer *Minnehaha* on Seton Lake, was to receive a half-interest in the Pioneer claim in return for putting the mill in operation and a payment of $500. Not long after, the agreement was changed to call for a one-stamp mill and the payment to a mere $150.[2]

Kinder, on the property much of the summer of 1901, did some tunnelling on the vein, and the one-stamp mill, little more than a toy, appears to have been installed at this time. Early in 1902, Allen and Atwood transferred the promised half-interest to Kinder. What use, if any, Kinder made of his mill is missing from the record, but the 1902 season must have been disappointing. At its end, he pledged his half-interest in the Pioneer claim as security for a $50 loan from Archie McDonald, the government road foreman.[3] Following this the property appears to have been inactive until late in the 1905 season, when Kinder built an arrastra and recovered gold worth $150 from three tons of ore before being shut down by frost.[4] From 1905 on, Kinder appears to have operated the Pioneer arrastra each summer and, in 1910, his operation was visited by Wm. Fleet Robertson, the provincial mineralogist:

> The development of the property is not great, but the conditions under which it is being operated are peculiar and worthy of special note. It is a "one-man mine," being owned and operated entirely by one man, F.H. Kinder, who is not a miner by trade, but who, single-handed, has successfully mined and milled enough ore each year to make a comfortable living.
>
> ... The owner has apparently done all the development and mining single-handed and alone; the ore has been mined, filled into sacks, and, where necessary, hoisted by hand and carried, either in a wheelbarrow or by the owner on his back, to a home-made arrastra ... capable of treating from 400 to 500 lb. of ore

Kinder's arrastra on Cadwallader Creek, 1910. Photo probably by Wm. Fleet Robertson. Public Archives of B.C. D-05723.

> a day. If wages can be made, and apparently they are, by such primitive methods and the total absence of capital, it speaks well for the gold-tenure of the quartz mined.
>
> ... The property has considerable merit, as a small mine, but does not promise to develop into a large one; the present output as it is being run would not exceed $600 during the season.[5]

The 1910 season was the last as a "one-man mine." On February 21, 1911, Kinder, as sole owner, sold the Pioneer claim to Andrew Ferguson and Frank Holten, both of Vancouver, for $20,000 just one day after purchasing the Allen and Atwood half-interest from a third party for $8,000. Judging from records available, Kinder appears to have managed to hang on to his half-interest despite defaulting on the $50 loan he had pledged it against in 1902. The Pioneer claim had been Crown-granted on March 6, 1901, and as such the appropriate place to record the loan transaction was the Land Registry in Kamloops rather than the Gold Commissioner's

The ruins of the arrastra in 1911. Photo by C. Camsell, Geological Survey of Canada, No. 16753.

office in Lillooet. In April 1903, Kinder, with the co-operation of Allen and a month later with the assignee of Allen's estate, obtained a conveyance of sorts for a half-interest that he recorded against the Pioneer title in the Land Registry.[6] In December 1905, Archie McDonald, stuck with a questionable claim against the Pioneer title, got his $50 back by selling his rights, whatever they were, to a third party.[7]

Arthur and Delina Noel were also involved in the Pioneer scheme, selling a block of adjacent claims to Ferguson and Holten. Starting early in 1908, the couple had been acquiring ground upstream or southeast of the Pioneer claim. Claims in good standing were purchased, usually for a few hundred dollars, and working from location posts rough surveys were run to determine if open ground lay between them. A gap between the Pioneer claim and those to the east was covered by the Union Jack fraction staked by Arthur Noel on April 23, 1908. The block, consisting of the Union Jack, Emmadale, East Pacific, Great Fox and Clifton claims, was sold to Ferguson and Holten on February 22, 1911, with the Corasand following in June after the Noels acquired an outstanding half-interest. None of the claims had good gold showings and the Countless (west of Pioneer), which did, was now in the hands of a new company developing the old Bend 'Or property. One report put the price of the Noels' package at $30,000 but erroneously includes the Pioneer claim, which if deducted, would leave the Noels' share at $10,000.[8] In addition, the Noels may well have retained an interest in the new venture.

Over the next few years, work on the property was done by a Pioneer syndicate consisting of Andrew and Peter Ferguson and Adolphus Williams, the latter holding an interest acquired from the estate of Frank Holten. Kinder's arrastra was abandoned and allowed to go to ruin; Pioneer's new owners were interested in proving up the tonnage needed for a full-scale mining operation. Small crews of up to five men were used, mostly in drifting on the No. 2 vein. Working east on the latter it was discovered that the Nos. 1 and 2 crossed about 70 feet west of the adit, the combined veins having a width of 10 to 12 feet for a 30-foot section and gold assays of just over half an ounce per ton. Overall, a modest tonnage of good ore was developed but, with vein widths generally a few feet or less and a mere 60 feet of rock forming the "backs"* above the workings, shaft* sinking would be needed to prove up the hoped-for tonnage. This was undertaken in the summer of 1914 and, to everyone's delight, the vein at a depth of 40 feet proved richer than on the levels above.[9]

In November 1915, Pioneer Gold Mines Limited, a million-share company, was incorporated to take over the property with three-quarters of the shares going to syndicate members for their holdings.[10] About 20 men were already at work installing a mill capable of handling 30 tons of ore a day. Transportation was no longer the problem it had been for the Bend 'Or in 1899. Shalalth was now served by the PGE railway and the new road from there was touted by an optimistic promoter as "one of the best in the Province, well maintained by the government, and suitable for automobile traffic from the railway to the mine."[11]

The mill, which started up in June 1916, was powered by a turbine wheel capable of developing 190 horsepower and was fed by an open flume, 1,700 feet long, that delivered water from Cadwallader Creek at a head of some 35 feet. The power system was unreliable, particularly in cold weather, when it became a never-ending battle to keep it going at all. Rather than using stamps in the mill, the ore was first put through a jaw crusher and then ground in a Chili mill, the latter sort of a machine-age arrastra with massive steel rollers mounted on horizontal arms that ground in a circular trough about a central axis. Mercury was used to produce an amalgam

Pioneer mine, 1916. Photo by C.W. Drysdale, Geological Survey of Canada, No. 38204.

that was recovered on an amalgamating table, and the tailings were run over a concentrating table before being discarded.[12]

Despite difficulty finding workers and supplies during World War I, the mine operated until the end of the 1919 season, when a mining company from eastern Canada took an option to purchase the property for $100,000. Over the four-year period 8,921 ounces of bullion worth $135,000 had been recovered. By now, sinking the inclined shaft had opened the mine to a depth of 187 feet below Kinder's original adit; extensive drifting had been done to test the vein both on 2 Level, 100 feet down, and on 3 Level at the bottom of the shaft. The new company had barely started work when their first option payment came due and, on asking for and being refused an extension of time, they threw up the option.[13]

Early in 1921 a Vancouver group, A.H. Wallbridge, A.E. Bull and others, acting on the advice of Charles Copp, formed a new Pioneer syndicate and leased the Pioneer property with an option to purchase. Over the next two years, under Copp's direction, a "small cheaply constructed leaching type of cyanide plant" was installed and the old workings cleaned out. "Breakages in machinery and other unforeseen contingencies occurred at intervals, hindering the work and hampering the management."[14] Following two years of bumbling Dave Sloan was called in to examine the property.

Sloan, a 41-year-old mining engineer, was a native of Perth, Ontario. Following graduation from Queen's University in 1905, he had followed the mining game from Arizona north to the Yukon and Alaska, including a number of years as assayer at the Surf Inlet gold mine on Princess Royal Island on the British Columbia coast. Then, troubled by long absences from a young family, he had dropped out to become an importer, only to be drawn back by the Pioneer assignment. Two weeks spent on sampling, assaying, panning and testing convinced him that the property had real possibilities, and in his report he stated that if the $30,000 required to start up could be raised, he was prepared to find $10,000 of it himself and take charge of the operation. Following two meetings with Sloan to discuss the proposal, the syndicate members, either unable or, more likely, unwilling to come up with more than $2,000, opted to sell the property.[15]

In the spring of 1924, Sloan was able to get a group from New York state to option Pioneer for $100,000 and to put up the $1,000 required to dewater the workings again. Two of the principals and their engineer, shaken by stories heard around Vancouver, took a superficial look at Pioneer, decided they had been flim-flammed, and demanded their $1,000 back on the grounds that the property had been misrepresented.

Discouraged by this setback, the syndicate approached Sloan with the suggestion that he take over Pioneer on a lease and option agreement and asked him to state his terms. A deal was struck for a five-year lease with a royalty of 15 percent of gross production to apply towards an option price of $100,000. In addition, the syndicate agreed to take a half-interest in the lease and to provide half of the $16,000 Sloan estimated it would cost to sink the shaft another 200 feet in an attempt to develop new ore. Sloan, in turn, divided his half-interest with an investor, J.I. Babe. The $16,000 was to be advanced in four monthly payments, two of them due before work got underway. However, there was a catch; old supplies on hand were to be taken over by the new group, leaving Sloan with only $3,000 in working capital after paying out $5,000 of the first $8,000. Most of the money went for a carload of dynamite, close to four years old and probably deteriorated to the point that it was dangerously touchy and apt to explode if jarred. By rights, it should have been destroyed but Sloan, desperately short of cash, had to chance using it.

Sloan took charge on July 21, 1924, and shaft sinking resumed soon after. This underway, an attempt was made to locate and mine some ore from old stopes* in the upper workings. Lady Luck smiled just in time and, by early September, Sloan had a gold brick as proof. As a later acquaintance recalled:

> But most of the $8000 . . . had been borrowed from the Bank of Montreal by the syndicate members. Andrew Lang, head of the bank in Vancouver, found he was financing a mine he had never heard of. There was no way to get hold of Dave [Sloan] at the remote mine but Dave could still write checks and overdrafts soon came in. Three or four months later Dave came to Vancouver carrying a paper parcel and was told to come to the bank at once. Lang used his best Scotch curses in telling Dave he was through. Dave said nothing but when Lang ran out of breath he unwrapped his parcel and handed Lang a gold brick worth about $3000. "Maybe this might help," Dave said. It did help and Lang said he could go on for another month if the overdrafts were not too big.[16]

The new ore came from an old stope or working just west of the shaft where earlier operators had stopped mining at an apparent pinch in the vein. Blasting an initial test round* revealed that the vein opened up again, carrying 3 to 4 feet of good ore that continued upwards. Two men, working part time, mined it while shaft sinking was underway. The ore was treated in the rickety old Chili mill, a relic from earlier days. About 75 percent of the gold content was recovered as amalgam which was then retorted and the residual sponge gold* melted down and cast into bricks. The latter were shipped to the government assay office and payment credited at the bank within four or five days from the time the cleanup got underway. In all, an estimated 800 tons were treated and bullion worth some $16,000 recovered with values of some $5 a ton left in the tailings.[17]

When cold weather forced a shutdown in November 1924, the shaft had been sunk an additional 142 feet to a new 4 Level and, on it, the vein drifted for two rounds in each direction. Both the grade and width of the ore were much the same as that on the upper levels. In the 1925 and 1926 seasons, mining continued on a "pay as you go" basis with bullion production of $61,000 in 1925 and nearly $100,000 the following year. The shaft was now down another 160 feet to 5 Level where the vein was up to five feet wide and the average value of ore, as calculated from an eight-day production run of development muck,* was over $50 a ton.[18] With values continuing, or even improving at depth, ample money in the bank and ore reserves with an estimated total value in excess of half a million dollars, there were plans for a much larger, more efficient operation. These involved a new, reliable power plant capable of operating year-round, a new vertical shaft and an up-to-date milling plant treating 100 tons of ore per day. Thanks to Dave Sloan, the Bridge River's reputation as "the valley of dead mines" was no more.

In the summer of 1927 the old flume along Cadwallader Creek was replaced by a wood-stave pipeline,* of 36-inch inside diameter and over a mile long, that delivered water to the mill at a head of 260 feet. The new system, utilizing waterwheels and belt-driven machinery, was capable of developing in excess of 500 horsepower, ample for the operation with the possible exception of late winter when low water in Cadwallader Creek or ice building up in the pipe reduced the flow.

At the same time, work was underway on both a second shaft and on a new mill. Number 2 Shaft, a vertical one with two compartments, was raised from 5 Level rather than sunk. It would not come through to the surface but instead was to be serviced from an adit driven into the hillside to a large chamber where the headframe and hoisting machinery were installed. In the new mill a cyanide* process would replace amalgamation with mercury. The ore would be ground in a cyanide solution capable of dissolving gold and silver and, after careful filtration, zinc dust added to the pregnant solution would precipitate the precious metals. Filtered and dried, the precipitate would be mixed with the necessary fluxes, smelted and the resulting bullion cast into bricks.

There were delays at the beginning of the 1928 season. Completion of the new adit to connect with No. 2 Shaft took longer than anticipated when heavy ground,* requiring extensive timbering, was encountered. This, plus a decision to sink the new shaft 500 feet below 5 Level, limited the amount of ore hoisted, and for just over two months, the newly completed cyanide mill had to be used to treat old tailings. The latter were found to have an average gold value of $4.10 a ton while, after treatment, the reworked tailings ran a mere $1.10 a ton. During the delay the old mill remained in operation, treating ore hoisted in the original No. 1 Shaft. As soon as the No. 2 Shaft was in full operation the old mill was permanently shut down and the new mill operated on a mixture of ore and old tailings. By now the crew was close to 50 men, most of them housed in a large bunkhouse built a year earlier. Other new buildings included a handful of small bungalows for married employees and a school for their children.[19]

In the four years of Sloan's management Pioneer had been transformed from the mine no one wanted for $100,000 to one able to turn out much more than that amount in bullion each year. Early in 1928, rather than waiting until royalty payments to the old company reached the $100,000 option price, the balance was paid in full and the lessees took over the property. Almost at once, it was turned over again, this time to a new public company, Pioneer Gold Mines of B.C. Limited, incorporated on March 29, 1928, with a capital of 2.5 million shares of $1 par value. In return, the group received 1.6 million shares of the new company, half going to the members of the Pioneer syndicate, the Wallbridge group, and the other half split between Sloan and his new partner, Colonel Victor Spencer, who had bought out J.I. Babe's interest some time earlier for a reported $40,000.[20] An additional 34,000 shares were reserved for key employees, provided they remained with Pioneer until the middle of 1930. Officers of the new company were: Colonel Victor Spencer, president; A.E. Bull, secretary-treasurer and vice-president; and David Sloan, managing director. The other directors, all members of the Pioneer syndicate, were Mrs. Helen Wallbridge, General J. Duff Stewart, Dr. R.B. Boucher, Dr. F.J. Nicholson and Dr. A.R. Thomson.

During 1929 the No. 2 Shaft was sunk the planned 500 feet to 9 Level and intermediate levels 6 to 8 cut at 125-foot intervals. The Main Vein, as it was now known, had been tested by a total of 850 feet of drifting on 7, 8 and 9 levels and found to be 4½ to 5½ feet in width and to have gold values in the $16.50 to $20 range, or slightly under one ounce of gold per ton. The mill operated for 360 days treated 26,760 tons, about half of it tailings from the old mill, and produced bullion worth just over $185,000.[21] In mid-December Pioneer advertised its good fortune by displaying three gold bricks worth $30,000 in the window of a jewelry store on Vancouver's Granville Street.

It was too good to last. On Christmas Eve 1929, water from Cadwallader Creek broke into the workings on 3 Level, poured down the shaft and flooded 8 and 9 levels of the mine.[22] Dave Sloan, contacted in Vancouver, ordered the small crew still at the mine to pull out and let it flood. There seemed to be no alternative as the company had no money and, with financial markets in turmoil at the beginning of the Great Depression, there was no prospect of raising any on short notice. Bob Eklof, one of the foremen, decided to stay on regardless; there was enough grub to last the winter and, with times so tough, little chance of his landing a job elsewhere. Most of the others did the same, and "borrowing" a pump from the Coronation property where Eklof had worked in 1925, they dragged it to Pioneer with horses and began pumping out the mine. There were new problems in January when the wood-stave pipeline carrying water to the turbines froze solid but, by February, the mine had been pumped out and mining resumed on a small scale. The crew had saved Pioneer from some heavy expenses, but even so, there were no pay cheques until months later when Pioneer's finances were in better shape.[23]

Despite the poor start, 1930 would turn out to be Pioneer's year. It began with known ore reserves above 5 Level virtually mined out but ended with close to a mile of drifting on the Main Vein on the new 6–9 levels below. Best of all, with the exception of the west drift on 6 Level, none of the workings had reached the limits of the ore zone. The Main Vein was narrower, averaging three feet, but rich enough that no attempt was made to hand sort and discard the barren wallrock broken in driving the 5½-foot-wide drifts. In all, some 29,000 tons of ore were milled that year, yielding bullion worth close to $286,000. Over 85 percent of the ore had come from the exploratory drifting with the remainder mined in three small stopes.[24] More development work would be needed to confirm it but there was every reason to expect the ore to be continuous between levels and that new stopes would keep the mill supplied for many years to come.

By now Sloan was planning other long-term developments. A new No. 3 Shaft would be raised from 9 Level to surface and sunk below that. Mill capacity would be almost tripled to 300 tons per day and work had already begun on a new hydroelectric plant on the Hurley River, some 3½ miles from the mine, capable of developing 750 horsepower. With the Hurley much larger than Cadwallader Creek it was hoped to avoid the water shortages that plagued the latter in the winter months.[25]

On April 1, 1931, the first day of their fourth year, Pioneer Gold Mines of B.C. confirmed their good fortune by paying an initial dividend of 3¢ per share. By now, the company was confident of paying that amount on a quarterly basis, calling for a total outlay of $210,000 a year, and still generating adequate funds for the planned developments. Estimated ore reserves at the mine had increased nine-fold from 25,000 to 225,000 tons although the grade had dropped slightly from 1.10 to 0.95 ounces of gold per ton.

At the company's annual meeting held in Vancouver on May 21, 1931, President Spencer held out the promise of even greater dividends once major projects were completed. He was just back from a successful trip to show off the mine to Ira B. Joralemon, a prominent American mining geologist, who was consulting for eastern interests holding Pioneer stock. Joralemon, much impressed by what he found, agreed with Sloan and others that the ore could be expected to go to great depth and that gold production would cover the cost of both the development work and the recently announced dividend rate. Alfred Bull, vice-president and secretary-treasurer, revealed that arrangements had been completed to purchase the Countless claim west of Pioneer, and that some underground workings, still in good ore, had already reached the Countless boundary. The cost, 80,000 Pioneer shares, was not disclosed at the meeting. Shareholders on their part were asked to approve the action of the board in authorizing the listing of Pioneer stock on the New York and Montreal curb markets.

At Pioneer mine, developments continued. By late 1932 two power plants on the Hurley River had been switched on and a third was planned. The No. 3 Shaft, now the main operating one, had been raised from 9 Level to surface and a 300-foot aerial tramway built to carry ore from the headframe to the mill. Below 9 Level the shaft had been sunk 625 feet to 14 Level with crosscuts* driven out to the Main Vein on intermediate levels 125 feet apart but, as yet, no drifting had been done on the vein itself.[26]

The Pioneer mine had arrived. In 1933, the year Al Abbott came to the Bridge River, it would mine and mill 100,519 tons of ore, producing 82,519 ounces of fine gold, valued at about $2,400,000 of which about $1,600,000 represented profits.[27]

Delina Noel with the first gold brick from the Lorne mine, 1916. Photo from W.A. Hutchings album.

# *Lorne Gold: An Honest-to-God Promotion, 1928–1931*

Dave Sloan's success at Pioneer drew others to the Bridge River. In March 1928, Lorne Gold Mines, Limited was incorporated to acquire and explore the ground between Pioneer and the Lorne, some 50 claims and fractions in all. It was a promoter's dream: two miles of favourable ground anchored at the Pioneer end by the Bend 'Or property, now known as the Coronation, and at the north end by the Lorne, both of which had already produced $100,000 or more in bullion.[1] Mining stocks were booming at the time and the Lorne was one of five British Columbia projects in the stable of Stobie, Forlong and Company, a major Toronto brokerage house.

Putting together the Lorne consolidation was a complex operation involving both companies and individuals, all expecting to be well rewarded for participating. Some holdings were acquired outright for cash or Lorne shares, while others were held under option with payments called for at some future date. It was made simpler and cheaper by the failure of all earlier attempts to develop a viable operation and by the perception that none of the known showings, with the possible exception of a new find on the Countless claim, seemed worthy of an immediate follow-up.

Promotion was the name of the game and greed the selling point. First, Stobie, Forlong had to gamble that they could recover their initial investment and, with luck, a great deal more in market action on Lorne stock. To achieve this, they, in turn, had to convince the public that buying Lorne stock was the highroad to fortune.

The consolidation completed, it was full speed ahead on a new low-level adit to test the veins of the Lorne property at depth. Stobie, Forlong, going all-out to generate interest in Lorne stock, would come tantalizingly close to coming up with a winner.

Turning back the clock, the story of the properties merged into the new Lorne scheme is, in many ways, the story of Arthur and Delina Noel. From 1900 on, the couple had been in and out of the area many times, either working on or acquiring interests in properties destined to be folded into Lorne Gold. Cash picked up along the way was ploughed back into the project at hand and when things turned sour the couple, nearly destitute, sometimes lived on wild game with their sole income being a little cash from the sale of a few furs.

In the 1909 season, Arthur Noel had leased the Bend 'Or property from bondholders who had taken over the company. Working with a crew of six men for five months, he rehabilitated both workings and mill, treating 240 tons of ore to recover bullion worth $2,298. Work continued in 1910, with the vein explored on three levels over a 200-foot vertical interval, while a fourth level started an additional 210 feet below had yet to reach the vein. That season another 135 tons were milled, but towards the end of the year the lease was terminated and the property sold.[2]

From 1911 to 1917, Coronation Mines, Limited, of Victoria, carried on exploration and a small amount of mining on their Coronation property, comprising the Bend 'Or ground and adjacent Countless claim. The work was under the direction of C.L. Copp. In 1926 a new company, Coronation Consolidated Mining Company, acquired the property, installed a new mill and in 1927 treated approximately 3,600 tons of ore yielding bullion worth $30,000.[3] Apparently it was not a financial success and the company deferred further mining to explore a possible extension of the Pioneer vein discovered on the Countless claim. In all, from 1899 to the 1927 shutdown, 12,297 tons of ore had been milled, and 7,052 ounces of gold and 1,004 of silver recovered.[4]

After leaving the Bend 'Or in 1910, the Noels, never far from the Bridge River, were back in 1916 as operators of the Lorne property at the north end of the favourable belt. The Lorne had a long history; all but one of the six claims involved had been staked in 1897 and arrastras operated in the early years. Between 1900 and 1901, the Lorne and adjacent Woodchuck groups had been bonded to an English syndicate for $225,000. During this period, close to a thousand feet of new underground work was completed and a five-stamp mill was installed on the Woodchuck claim. Gold values found in the new workings were much lower than those in older, near-surface workings and the bond was dropped in September 1901. The mill, not quite completed, was never operated. The claim owners, already involved in litigation with the syndicate, fell to squabbling among themselves over ownership of the mill and the right to use it. Nothing was resolved, and the disgruntled claim owners went back to operating arrastras. In 1910 the Lorne and Woodchuck groups, of three claims each, were taken over by Lorne Amalgamated Mines, Limited. The mill, completed at last, operated part of that season and for brief periods in the following two seasons.

In the spring of 1916, the Noels, apparently holding an operating agreement with Lorne Amalgamated, freighted in heavy loads of equipment and supplies over the government wagon road from the PGE railhead at Shalalth. Underground workings were rehabilitated, the mill overhauled and new buildings added to a spruced-up campsite in the centre of a ten-acre clearing. One new building, a bit away from the rest, was a substantial log bungalow about 20 by 40 feet:

> Mr. and Mrs. Noel intend to spend the winter here and have just built a very comfortable home in which they have probably one of the most unique fireplaces in the world. It is built of rock taken from the mine and gold nuggets are to be seen sticking out of the rock in many places. Besides being very artistic, it will give a lot of comfort to the owners during the long winter.[5]

When the Noels took over, both surface and underground work had been done on showings spread across a thousand feet of hillside. From north to south they were referred to as the Wedge, King and Woodchuck lodes or veins but, as yet, there was no telling whether they were strong, continuous veins or simply little teasers that would never amount to much. That first season the eight-man crew drove 155 feet of crosscuts and drifts, much of the footage on the King No. 4 Adit, driven from a point just above the mill and intended to cut the King Vein some 250 feet below its surface showing. A raise* was begun from it to the upper levels which, on

Freighting supplies to the Lorne property, 1916. Photo from W.A. Hutchings album.

Lorne mining camp, 1916. Photo from W.A. Hutchings album.

completion, would halve the cost of moving ore to the mill. Some 1,000 tons of ore were milled yielding bullion worth $11,324.32. Work carried on in the 1917 season and, although no stoping was done, 1,500 tons of ore obtained in driving 532 feet of underground workings yielded $22,500. Things slowed in 1918, due to the manpower shortage in World War I. Only a minor amount of development work was done, and a mere 380 tons of ore was milled.[6]

The Noels stuck with it but, even with experienced miners available again, the mine never bounced back. By the 1923 season the Noels, alone on the property, were doing the development work themselves. They kept at it until 1925, trenching veins on surface and attempting to cut them in underground workings. Results varied; a few samples taken from a narrow section of one vein assayed $100 a ton but, more often than not, the assays were discouragingly low at less than $10 a ton. There was still hope and, in 1925, a newly appointed government mining engineer wrote of his first visit to the Lorne:

> From a brief examination the opinion was formed that the economic possibilities are greater than might be gathered from the history of the mine and from the general tenor of references

> that have been made to it. The extent and disconnected character of the many workings has had the effect of creating more or less vague impressions of the value of the ore-deposits, while the number of veins which have been prospected and their varying strikes* have still further increased the difficulty of considering the property as a workable unit.[7]

The Noels' attempt to develop the Lorne ended in 1926 when the Lorne Amalgamated company, revived by court order, optioned the property to two American promoters.[8] The company, drifting for years, had been struck from the register of companies for failure to file annual reports after 1921 and the claims, put up for tax sale, were acquired and held in trust by H.H. Shandley, the company's lawyer. A deal made with Arthur and Delina Noel gave each an eighth interest in Lorne Amalgamated, presumably for their work on the property and abandonment of any further claims against the company.

The couple left the Lorne at this time. For Arthur Noel, now almost 60, what would prove to be his final attempt to make a mine in the Bridge River had ended in failure, although he would later play an important role in Stobie, Forlong's Lorne consolidation and dabble in other mining

The blacksmith's shop and assay office at the Lorne camp, 1916. Photo from W.A. Hutchings album.

Arthur Noel with amalgam from the first cleanup, 1916. Photo from W.A. Hutchings album.

ventures. By 1929 the couple had separated, and Delina, bitten by the mining bug, would return to spend many more summers in the area, her base the Lorne cabin built for the couple in 1916.

Who or perhaps what drew Stobie, Forlong's attention to the Lorne scheme is unknown. One possibility is R.B. Lamb, a mining engineer and financier. Born in Ballarat, Australia in 1876 and the son of a prospector, he had followed the mining game in Australia and throughout North and

Central America, most recently in British Columbia and Ontario. Home was where he hung his hat that day and, between 1928 and 1930, addresses are listed in Toronto, Vancouver and San Francisco. Lamb, either acting on his own or as Stobie, Forlong's agent, had negotiated the purchase of an outstanding option on the six-claim Lorne group, the key to the consolidation.[9]

The option, calling for cash payments totalling $150,000, had been granted in 1926 to two American promoters, D.S. Beals and H.E. Hopper, by Lorne Amalgamated Mines, Limited. Its terms called for the final payment to be made by May 1, 1929, and, in addition, there was an undertaking to complete 800 feet of underground work by March 30, 1927. Some of the scheduled cash payments had been made but the property lay idle through 1926 and 1927 and, in April 1927, the option agreement, technically in default, had been modified to drop the work commitment. A further change, in February 1928, extended the date for the final payment to January 30, 1931.[10] Prior to this date, the promoters had transferred the option to their own company, Lorne Mining and Milling Company, of Seattle. That company and the promoters, in negotiations with Lamb, agreed to transfer the option to Stobie, Forlong in return for 250,000 shares of the new Lorne Gold company and $50,000 in cash. Next, Stobie, Forlong, in effect dealing with themselves, wrote up a deal with their own Lorne Gold company agreeing to transfer the option once more in return for 550,000 shares of the latter. No cash was involved, and presumably Lorne Gold was expected to come up with the $105,000 still due in option payments and the $50,000 for Lorne Mining and Milling Company. Then in early April 1928, 200,000 of Stobie, Forlong's 550,000 shares were transferred to R.B. Lamb and a further 150,000 to six individuals involved in other aspects of the consolidation.

In contrast, the deal for the other important property, the Coronation, had been straightforward. Coronation Consolidated's promoters, badly burned by their attempt at mining, were probably glad to see the property go for 600,000 Lorne Gold shares, although they did have the good sense to hang onto the Countless claim. Deals on the remaining properties were smaller, involving either cash, shares or a combination of the two. In most of the deals a portion of the payout went to Arthur Noel, usually in the form of shares. When all was done, Lorne Gold controlled the ground they wanted, provided, of course, that they could come up with the cash still due under the Lorne option. In all, 1,475,000 of Lorne Gold's authorized capital of 3.5 million shares of $1 par value were issued for the properties.[11]

## Where Big Developments are Occurring

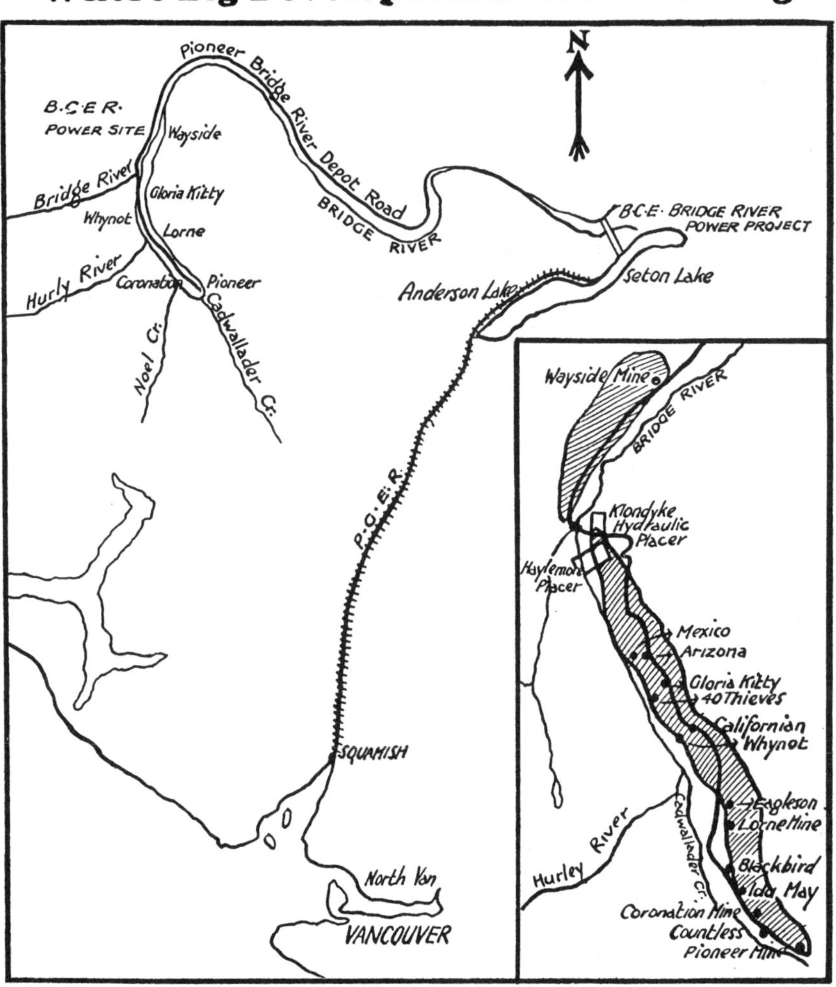


Map showing the relation of the Bridge River mining field, and the Bridge River power development project of the British Columbia Electric Railway Company to the Provincially-owned Pacific Great Eastern Railway, suggesting a possible development of hydro-electric energy and mineral wealth which may alter the outlook for that road. The shaded portion indicates approximately the upthrust of the diorite formation in which occur the free gold ores.

*We shall be pleased to furnish further details on request.*

**STOBIE·FORLONG & Co**

Members:
Standard Stock and Mining Exchange, Toronto, Vancouver Stock Exchange and other Exchanges.

VANCOUVER OFFICE:
35 COMMERCE BUILDING
Seymour 9114-5-6

Head Office:
TORONTO

PRIVATE WIRE SYSTEM CONNECTING PRINCIPAL CANADIAN CENTRES.

A 1928 promotional advertisement, one of several such issued by Stobie Forlong & Co.

Lorne Gold's main project, started in May 1928, was a low-level adit to explore the veins on the Lorne group at a depth of 800 to 900 feet below the surface, some 630 feet beneath the earlier workings. It was a gamble, going "full speed ahead with no expense spared" the antithesis of the "pay as you go" approach that had been so successful at Pioneer. Stobie, Forlong's press releases and handouts linked Pioneer and Lorne together, referring to the former's production in one sentence and the latter's potential in the next, a skillful mixture of fact and fantasy. Then, to top it off, came the pitch that the quartz veins, with their high-grade* ore shoots marked by ribbon structure, were similar to those found in the Motherlode district of California and, like them, certain to persist to great depth. The would-be investor was offered "the opportunity to become early identified with an enterprise of great promise."[12]

Work on the Lorne low-level adit was done well despite the rush and lack of careful planning. B.W.W. McDougall, consulting engineer, and E.P. Crawford, mine manager, must have had their hands full keeping the crew of up to 90 men busy and out of one another's way. A bunkhouse and cookhouse, both two-storey and 30 by 110 feet, plus a number of smaller structures were built on a bench above the portal of the new adit. The old Lorne minesite higher up the slope was abandoned. Initially water to power air compressors and a sawmill came from Cadwallader Creek after cleaning out and repairing 1½ miles of old ditch and flume line, laying 2,000 feet of wood-stave pipe and building a diversion dam and headgate. The first 100 feet of the new adit was driven by hand but, with the air compressor in operation, three crews working around the clock advanced the heading* at a rate of 20 feet a day. In addition to the usual air drills, they were supplied with a mechanical mucking machine and a battery-operated mine locomotive to speed the work. By the end of 1928, the adit, straight as a die and 1,693 feet long, had entered the block of ground between the overlying King and Woodchuck workings.[13] Two or three small quartz veins, one of them showing free gold, had already been cut but, as yet, no attempt was made to explore them. The speed paid off, with the promoters, wise in the ways of publicity, making the most of it in a blizzard of press reports.

During 1929, 5,054 feet of drifting and crosscutting and 264 feet of raising were completed. From the main adit, now in 2,390 feet, a long crosscut had been driven in a northerly direction from near the 1,750 foot mark to test the King Vein and the ground beyond it, while from the opposite side of the adit another short crosscut led to beneath the Woodchuck Vein workings.

The King Vein, intersected at 518 feet in the long crosscut, showed the most promise. It had been drifted for 150 feet to the west, where it was cut off by a strong fault,* and for 700 feet to the east, where an oreshoot roughly 450 feet long with an average width of three feet and values around $20 a ton had been opened up. An inclined raise to connect with the old King Vein workings above had been started from the crosscut. Three other veins were explored by drifts run from the crosscuts but, aside from a few good assays, there was nothing to get excited over.[14]

Press releases kept coming all year, most of them based on reports by McDougall or N.H. Atkinson, who had replaced Crawford as mine superintendent. There were hints of a mill; then late in 1929, the directors announced that McDougall had been instructed to seek tenders for a mill and hydroelectric plant.[15]

It was all whistling past the graveyard. Stobie, Forlong and Company, caught up in an investigation of the market for mining stocks, declared bankruptcy on January 30, 1930, cutting off all funding for the Lorne project. Lorne Gold, in debt and with payments coming due, would be lucky to avoid the same fate.

The mining brokers' troubles had begun on November 7, 1929, when *The Financial Post* of Toronto kicked off a series of articles with the headline, "Gigantic Manipulation in Mining Industry Hits Public Confidence." If not true at the time it certainly was when the series concluded nine weeks later. By then, in addition to the exposé, there was growing awareness that Black Thursday, October 24, 1929 - the first day of panic selling on the New York Stock Exchange - might be much more than a long-overdue correction on that market and, as such, totally unrelated to business conditions in Canada.

The first article made no specific charges but, in a mangled metaphor, had the brokers "rolling in clover - and big new motor cars" then, a few short one-sentence paragraphs later, transformed them into "the pest, the worm and the parasite" attacking the "apples" in the "mining orchard."

Later there were many charges. The most serious, made in the second article, was that some brokers were "bucketing" or playing the market against their customers. It involved a technique, refined somewhat from cruder methods used in the past, of placing a customer's order for stock on Toronto's Standard Stock and Mining Exchange and, about the same time, placing an offsetting order to sell short a portion of the purchase. Thus, at any given time, a broker engaged in "bucketing" might actually hold as few as one-third of the shares shown on their customers' accounts. One such broker, invited to express his viewpoint, defended the practice, claiming it provided "facilities for a much larger degree of... speculation than would otherwise be possible," helped maintain stability in the market and, if that were not enough, much of the money his firm had made had been ploughed back into developing mining properties without asking the public to put up a nickel.[16]

There were other tricks too. Many stocks were bought on margin, with the client putting up perhaps a third of the purchase price, and the brokers saw nothing wrong with charging interest on the outstanding amount of the fictitious purchase. Then there was "wash trading," in which stocks were swapped back and forth among brokers to create the impression of heavy trading and to manipulate the price and, perhaps the lowest of all the tricks, gouging on transactions by showing purchases made at the high price for the day and sales at the low point.

It was a lively subject and *The Financial Post* printed a host of letters to the editor describing experiences with brokers or calling for general reform. The first move from government came on January 11, 1930, two days after the final article, when the Attorney-General of Alberta ordered the arrest of two principals of one of the largest firms. The Ontario government followed suit and, in an early morning raid on January 30, 1930, 12 men connected with five other prominent brokerages were picked up by police, brought to Toronto City Hall and charged with conspiring "by deceit or falsehood or other fraudulent means to defraud the public."[17] Stobie, Forlong was one of the firms involved. Its principals, Malcolm Stobie and C.J. Forlong, were released on bail of $100,000 each, half of it posted by General D.M. Hogarth, later a director of Pioneer. The company's books and securities were seized by police, and employees were sent home at noon the same day when the firm's lawyers filed a petition in bankruptcy and an associated company went into voluntary liquidation.

At a week-long trial in early November 1930, the government-appointed accountant charged with sorting out Stobie, Forlong's operation testified that on the day of the arrest the firm was short close to $4,000,000 worth of a mere five of the many stocks they dealt in. The jury, unimpressed by the arguments of the defendants' high-priced legal talent, found Stobie and Forlong "guilty of charges to defraud the public and conspiracy to affect the price of a stock" and the pair were sentenced to three and 2½ years in jail respectively.[18] Their appeals, together with four others, were dismissed on March 18, 1931, and, in an unusual move, judgment was handed down without the customary advance notice. Perhaps there were suspicions of attempts to skip the country. In any case the police moved swiftly and all six were rounded up and brought to the Toronto sheriff's office by 2:00 that afternoon. Joined by about 40 of their friends they chatted and smoked, the gathering having "many of the incidents of a send-off at a railway station."[19] The six were then shuttled to the Toronto jail in two large cars, their carefree attitude gone as they entered the doors of the first way-station on their journey to the Kingston penitentiary.

With the Stobie, Forlong financing gone, Lorne Gold's project was doomed without new money, provided either by shareholders themselves or by striking a deal with a well-heeled third party. In all, $350,000 worth of work had resulted in the discovery of 60,000 tons of ore with a total value close to the cash outlay. Initially, there had been doubters over the wisdom of driving the low-level adit but, despite this, there was general agreement that the work had been done well with good value for the amount spent. The next stage, if there was one, was going to be an expensive gamble. Who could have guessed that a bonanza lay within 100 feet of the point where drifting on one of the smaller veins had been halted on striking a major fault!

At Lorne Gold's annual meeting, held in Vancouver on May 1, 1930, the shareholders approved a refinancing scheme calling for creation of 500,000 preference shares of $1 par value carrying a 7 percent cumulative dividend. The directors proposed to offer 400,000 units, consisting of one preference share and one common share, to the shareholders at 85¢ each. If all were taken up the company could pay off the $97,000 still owing in property payments and have almost enough left to install a mill, estimated to cost $150,000.[20] Six directors were elected for the coming year, all but one from Vancouver, the exception being D.S. Beals, the Seattle promoter. F.W. Rounsefell, a financier, was the new president, and Arthur Noel, with a much-reduced but still important shareholding, joined the board for the first time.

There was no easy answer to Lorne Gold's problems. Shareholders had approved refinancing but apparently had second thoughts when it came to putting up money and the scheme collapsed. Premier Gold Mining Company, operators of a profitable mine near Stewart, B.C., expressed interest in the Lorne property, but after an examination in the summer of 1930, decided that it was not for them.[21]

In mid-January 1931, F.W. Rounsefell mailed a desperate appeal to the shareholders reminding them that they stood to lose the Lorne group, on which nearly all the work had been done, unless a final payment of $95,000 was made before January 30, 1931. In it, he quoted from a letter by H.H. Shandley, representing Lorne Amalgamated Mines, Limited:

> To summarize (1) There will be no extension of time granted for the payment of the balance of the money. (2) If any party tenders $95,000 to me I am going to take it and give them title to the property subject to the agreement for sale.[22]

Two rescue proposals were presented to an extraordinary meeting of Lorne Gold shareholders held on February 4, 1931. The first, from Bralco Development and Investment Company, Limited, a private company, was to form a new company in which the present Lorne shareholders would retain an interest. The second, from Toronto General Trusts on behalf of undisclosed clients, was to advance $97,000 in the form of a mortgage in return for a $50,000 bonus and the right to purchase Lorne shares at graduated prices. Bralco's proposal was the choice of the majority of the shares represented at the meeting but it was withdrawn when Bralco's lawyers pointed out that Lorne's articles of association called for one vote per shareholder, regardless of the number of shares held, and, on that basis, the Toronto General Trusts offer was the winning one.[23]

A second extraordinary meeting was called for March 24, 1931, to consider amending Lorne's articles of association to give one vote per share and to consider a new offer for the property from its neighbor, Pioneer Gold Mines of B.C. Limited. Pioneer offered to advance the $95,000 now past due, to install the old Pioneer mill on the property and to provide working capital in return for a 55 percent interest in a new operating company. Neither of the previous proposals had been resubmitted.

Bralco, anticipating an offer from Pioneer, may well have held back in order to get a look at it; in any case when the Lorne meeting was called to order there was a new Bralco proposal to consider as well. In return for a 60 percent interest in the new company, Bralco offered to advance money to pay off the existing obligations, plus $50,000 for development work and $150,000 for a mill and power plant: in all some $300,000.

Shareholders represented at the meeting had their own ideas and the proposed one vote per share was turned down. In the confusion that followed the meeting was adjourned for two days, and when it reconvened the shareholders, each with their single vote, considered the two proposals. This time, Bralco won hands down, 502 to 38 and, with the offer accepted, the next phase of development began.[24]

The Bralorne mill, 1934. Photo by Leonard Frank, Vancouver Public Library VPL 14851.

# *Bralorne: Austin Taylor Takes Control, 1931–1932*

Bralorne Mines Limited, incorporated April 22, 1931, as the new operating company took over work on the Lorne property. Its authorized capital was one million shares of no par value and all would be issued, 60 percent to Bralco and 40 percent to Lorne Gold Mines, Limited. Bralco had taken on a $300,000 gamble, as witnessed by the fact that a number of major mining companies had already backed away from the project, but there was a reasonable chance of recovering the money advanced and, if more ore were found, it could be a winner.

Bralco Development and Investment Company, Limited, its full title, was a private company. Its principal was Austin C. Taylor, a Vancouver financier and industrialist, holding one share short of 50 percent of the issued shares while the remainder were held by associates: Neil McQueen and W.W. Boultbee, each with 20 percent, H.G. Fowler with 10 percent and the single share by an employee of the incorporating law firm.[1]

Taylor, born in Toronto in 1889, had been commissioned a major in the Canadian Army in World War I and, in November 1917, was sent to British Columbia with orders to take charge of and speed up the production of spruce lumber, urgently needed for aircraft production. Taylor knew next to nothing about spruce; at the time he was directing artillery shell production at Montreal Locomotive Works, where he had made his reputation as a man who could organize and get things done in a hurry. He showed the same talents in his new job and spruce production was soon at record levels. Six feet tall and with an athletic build, he was described by one writer as "the personification of aggressive force."[2] Not a man to be trifled with, he knew what he wanted and usually got it, at times intimidating others with his habit of lowering his head and glaring at them through bushy eyebrows. Remaining in the west following the war, Taylor had been involved in a number of business ventures, including the formation of Home Oil Distributors, a major retailer. Now, through Bralorne, he was joining the mining game.

Taylor must have used technical help in preparing the bid on the Lorne property, but, surprisingly, there was no group of key personnel standing by ready to take over once the transfer was completed. For the moment, H.C. "Pop" Wilmot, a metallurgist who had been at the Coronation property in 1928, was in charge and he may well have been the one to recommend the Lorne project to Taylor.[3] Neil McQueen, Taylor's associate in Bralco, was a geologist and early in the summer of 1931 McQueen and his wife took up residence at Bralorne.

In August 1931, Ira Joralemon, a mining consultant based in San Francisco, was brought in to evaluate the property. He was already familiar with the Bridge River, having examined and reported on Pioneer a few months earlier and, on his way home, met with Austin Taylor in Vancouver and taken on the new assignment. Joralemon, then in his late forties, had already had a remarkable career involving property examinations and exploration on all the continents save Australia. Initially employed by mining companies, he had struck out on his own as a consultant in 1922, following service in World War I and a few years spent operating mines rather than in finding them.

After spending nine days examining and sampling underground workings and surface showings, Joralemon was convinced that Taylor's Bralorne gamble would pay off and, beyond that, there was a good chance of developing a multi-million dollar mine. The immediate need was for a good manager, Joralemon considering Wilmot "entirely ignorant of the art of running a small, high grade mine,"[4] although he would later concede that the man was a good metallurgist. To replace him Joralemon recommended R. "Dick" Bosustow, manager of the soon-to-be shut down Presidio mine at Shafter, Texas.

Bosustow, a "Cousin Jack*" or Cornishman, about 40 at the time, had emigrated to the United States as a young man and taken out citizenship. Before he could be hired, Canadian immigration authorities would have to be convinced that no Canadian qualified to fill the job was available. Assured that Bosustow was interested in coming to Bralorne, Joralemon drafted a "help wanted" advertisement that was almost a biography of Bosustow to be run in Canadian newspapers. Despite a shortage of jobs, there were no replies from Canadians and Bosustow, clearing the immigration hurdle, arrived at Bralorne in mid-November 1931.

Others joining the Bralorne organization in 1931 and destined to become key personnel included: John Muir, master mechanic; E.J. "Ted" Chenoweth, another Cousin Jack and later mine superintendent; D.N. "Don" Matheson as mine engineer and, in time, Busustow's successor; A.D. Hotson as electrician; and A.A. "Ad" Almstrom as assayer and later

The Bralorne Camp, 1934, with the original Lorne workings high on the hillside to the right. Photo by Leonard Frank, Vancouver Public Library VPL 10040.

mill superintendent. Most were hired before Joralemon's visit in August and all but Almstrom before Bosustow's arrival, making it unlikely that either man was involved in their selection.

Bosustow, a short, somewhat heavy-set man with piercing bright eyes, was the right man for the job. Extremely competent technically, he seemed to be on top of everything going on at Bralorne and always on the lookout for better ways of doing things. Some employees found him reserved, even crusty, but when circulating through both mine and mill he would stop and talk to them about their work. He seemed to have almost total recall of previous conversations, with one failing, often joked about: an inability to remember the name of the person he was talking to. One young staff member on arrival at Bralorne was told by Bosustow that he was a liability until he had been taught some things, the most important being to learn to get along with the men and, for that reason, he would be starting out in the bunkhouse. Later, if he succeeded at this, there might be a place for him in the staff house.[5]

During 1931 a total of 2,000 feet of underground work was done on the King Vein, consisting of 1,400 feet of drifting, 500 feet of raising to complete the raise from the new low-level adit to the surface, and 100 feet of sinking on a new internal shaft to explore the King Vein at depth. A second-hand, 100-ton-per-day flotation* mill had been purchased and moved from the Dunwell property near Stewart, B.C., to Bralorne. Operated for a mere eight months, most of the machinery was as good as new and the mill itself was considered one of the most efficient in the province. In addition, a hydroelectric plant and some mill machinery came from the Cork-Province property near Kaslo, B.C. By the latter part of the year over 100 men were at work, the majority of them on the new mill.

Everything went without a hitch and the new power plant and mill went into operation on February 6, 1932. Unlike Pioneer, Bralorne did not use a cyanide plant to recover the gold but rather amalgamation, the old standby, followed by flotation to recover a lower grade concentrate, mainly sulphide minerals, that was hand sacked and shipped to the Tacoma smelter for treatment. In all probability the decision was influenced by cost; second-hand machinery was available and, unless new ore was found, Bralorne would be lucky to do more than recover the initial investment. Shortly before the mill started up, Bosustow brought in Fred Gray, a fellow American, as his mill superintendent. Over the next few years, Gray would make many innovative changes to improve performance, including replacing amalgamating plates with blanket tables* and the introduction of specially designed jigs* that yielded a heavy, gold-rich concentrate for amalgamation.[6] The unusual mill, enlarged and somewhat modified, served Bralorne's needs for close to 30 years until finally replaced by a modern cyanide plant.

About this time, Bralorne brought in the Bridge River's first medical doctor, Donald King, under an arrangement that made his services available to Bralorne, Pioneer and B.R.X. employees and their dependents for a charge of $1 a month for a single man and $2 for a family.[7] To King, a young University of Toronto graduate interning at the Vancouver General Hospital and struggling to pay off university debts, the salary of $250 a month seemed too good to be true.

Lorne Gold's muddled finances were finally getting straightened out and its directors called a meeting for August 5, 1932, to consider winding up the company. The total cost of settling outstanding debts and the wind-up was put at $44,500, but not much cash was involved, because the two principal creditors, Stobie, Forlong Assets and Bralco, both agreed to accept Lorne Gold shares at 10¢ a share in payment. In addition, Lorne shareholders would be given the opportunity to purchase more shares at 10¢ on a pro rata basis. Once everything was taken care of, the directors estimated that the 400,000 Bralorne shares the company held could be distributed on the basis of one Bralorne share for every eight Lorne shares held.

Notice of Lorne's meeting was accompanied by a letter from W.W. Boultbee, secretary-treasurer of Bralorne, reporting on developments at the mine, mostly good news but with some disturbing items. Between February 8 and May 31, 1932, the mill treated 11,119 tons of ore, well in excess of the rated 100 tons per day, and recovered bullion with a total value of $157,475. Mill efficiency was improving too. For the entire period, the tailings had averaged gold values of $1.12 per ton but, by the month of May, this had dropped to 82¢ a ton. The inclined shaft sunk from the low-level adit, now referred to as 8 Level, was down 260 feet and drifting had been done on the King Vein on 9 Level, 110 feet below and on the 10 Level at the bottom of the shaft. Results were mixed; good on 9 Level and disappointing on 10 Level where 200 feet of drifting had been completed. In addition there was some drifting on the King Vein on 4 Level, as the lowest of the old Lorne workings was now referred to, and a little stoping done on a small shoot of high-grade ore found on the Shaft Vein on 8 Level. The known ore reserves would keep the mill going for another seven months and there were hopes of adding more ore through the exploration program now underway.[8]

At the meeting on August 5, Lorne Gold shareholders voted unanimously to liquidate their company and distribute the Bralorne shares. The meeting, held at 11:00 a.m., lasted less than half an hour and, as soon as the results were known, Lorne shares began to rise, with 10,400 shares changing hands at 9¢ before the market closed for the day.[9]

On June 1, 1933, Bralorne shareholders were mailed the annual report for the year ended December 31, 1932, together with a notice of the annual meeting to be held on the fifteenth of the month. Reporting on the year, R. Bosustow, mine manager, stated that over 4,000 feet of exploration work had been done, almost all of it on the King Vein, and that a new orebody found to the east on the vein was known to extend from 8 Level to well above 4 Level, a vertical distance of more than 500 feet. Thanks to this find, he put a conservative estimate of reserves at 30,000 tons, an increase of 5,000 tons over the year despite the milling of some 33,000 tons. But the exciting news came in Bosustow's supplementary report, dated May 1, 1933, on subsequent developments: an exceptional orebody, estimated at 125,000 tons and with a potential for much more, had been found west of the fault that cut off the King Vein in the old Lorne workings![10]

At the meeting, Bralorne president Austin Taylor announced that a new power plant would be installed within the next two months, a new crusher was on order, and when everything was installed by late in the year the mill capacity would be 250 tons per day. Secretary-treasurer Boultbee read a statement to the effect that only $30,000 was still outstanding on the $305,000-odd that Bralco had advanced to Bralorne but, with $80,000 in the latter's treasury, that could be paid off at any time.[11]

Everything changed with the discovery of ore west of the newly-named No. 1 Fault. Now Bralorne had a future and, with an assured supply of ore, there could be long-term planning instead of a day-to-day struggle to keep the mill operating at 100 tons per day, well below actual capacity. Either through brilliant deduction or blind luck, the right place had been chosen for the drive through the fault and success had come almost at once. On 8 Level, the old Lorne workings had reached the fault in two places, the first along the King Vein and the second 200 feet to the north in a drift along a narrow vein believed to be the Wedge Vein. Both veins were cut off abruptly at the fault face and there was no clear indication whether the drift, once through the fault, should be turned to the right or the left to explore for the displaced segment of the vein. Rather than continuing the drift on the King Vein through the fault, the decision was to extend the more northerly drift on the Wedge Vein on the chance that the King Vein, displaced along the fault, lay close at hand. Sixty feet beyond Lorne's last round, they hit the jackpot!

Either Colonel H.H. Yuill or Ira Joralemon may have come up with the idea of extending the drift on the Wedge Vein and, as is often the case with an important discovery, it is not clear where credit belongs. A mining journal published a few months after the discovery stated that the recommendation came from Yuill, a consulting engineer who had an office in Bralorne and presumably did some work for the company, but Joralemon, in his autobiography, implies it was his idea.[12] Perhaps it should be shared. Although Joralemon had not visited Bralorne in the six months prior to the find, it seems likely that the two men would have discussed exploration west of the fault on Joralemon's earlier visits in August 1931 and June 1932. Then, there is always another possibility: simple, blind luck. A woman living at Bralorne at the time wrote, years later, that it had been found when the drift was extended a few feet for use as a latrine.[13] Whatever the true story, Joralemon's best find and his alone came a few years later when his persistence in exploring the nearby Bradian property paid off.

During 1933, the King Vein lying west of No. 1 Fault was explored on 6, 7, 8 and 10 levels and the inclined shaft sunk below 10 Level. The new find would turn out to be a remarkable orebody, unlike any found in the Bridge River either before or since. The best ground, referred to as the A Block, lay in a wedge between two north-trending faults, No. 1, dipping roughly 70° to the east and No. 2, 60° west. The apex of the wedge where the two faults ran together lay above 6 Level while on 8 Level, 250 feet below, the faults were about 450 feet apart. Apparently the faults were active when the gold-bearing quartz was being emplaced, as an important new C Vein was found along the continuation of No. 2 Fault, lying nearly at right angles to the King Vein. To the west the D Block, a displaced segment of the King vein, lay between No. 2 and No. 3 Faults.

In the A Block the greatest change was in the King Vein itself, with many more branches and a mining width much greater than in the earlier workings. On 7 Level, the uppermost, with a continuous length of the King Vein, the two faults lay 270 feet apart but the vein, bent and twisted as if squeezed in a vise, had a length of 400 feet and an average mining width of 12 feet. A short distance below the level, strands of gold-bearing quartz were so numerous and so close together that a block of ore, 60 feet wide, had an average value of 0.5 ounces of gold per ton. Below 7 Level, the distance between Nos. 1 and 2 faults increased and, although the bending of the vein was not as pronounced, its average width was still much greater than in the Lorne Block, east of No. 1 Fault.[14]

The new find was a bonanza. Bralorne had arrived. Ore reserves at the end of 1933 were reported as 230,000 tons with an average grade of 0.6 ounces of gold per ton and on April 16, 1934, the company paid an initial dividend of 12½¢ a share.[15]

Stoping on the King Vein, Bralorne Mine, 1934. The whitish, ribboned material above the miners is largely quartz, which carries the gold. Photo by Leonard Frank, Vancouver Public Library VPL 14847.

# *Great Years at the Mines*
## *1933–1941*

Bill Dunn, dressed for work underground, 1935.
Photo courtesy of W. Dunn.

# *Working in the Mines*

1933 was the first of the boom years. Pioneer, prospering, had been a dividend payer for two years and Bralorne, with its new-found bonanza in the King Mine, would soon be joining it. Others, both companies and individuals, rushed to get in on the action and during 1933 about 4,000 claims were staked, blanketing 20 miles of favourable ground from McGillivray Pass north to Tyaughton Lake. To the government mining inspector it was a rerun of earlier rushes, "on a high percentage of [the claims], the outstanding features are the location-posts" although he did go on to concede that of the over thirty companies and operators involved the majority "were honestly endeavouring to find and develop something worthwhile."[1] Ironically, despite promoters' predictions and a great deal of exploration work only two other properties, Minto and Wayside, have produced significant amounts of gold to date.

The activity drew men to the Bridge River. Some sought jobs, others business opportunities and still others, the prospectors, to find their Eldoradoes. Initially, many found work with the new companies, now flush with cash and impatient to explore their properties. Dreams of important new mines faded over the next few years and, more and more, men looked to Bralorne and Pioneer for jobs. The two were never large employers: by 1935 Bralorne had a crew of 320 and Pioneer 280, but increasingly it was their cash, paid out either in wages or for supplies and services, that carried the Bridge River's economy.

The Bridge River was still remote and, despite the boom, nothing had been done to lessen the isolation. The 15-mile gap, bridged by the PGE's rail shuttle and ferries on Seton Lake, still separated its roads from those in the rest of the province. Shalalth, the starting point for the road trip over Mission Mountain, was a 9½-hour boat and train ride from Vancouver and, with luck, about the same if one travelled the road route through the Fraser Canyon.

## *Finding a Job*

Bill Dunn was 18 when he made his way to the Bridge River and joined "the line" at Bralorne in April 1935. Every morning up to 50 men would stand waiting as the shift went underground, each hoping to be the one chosen if Ted Chenoweth, the mine superintendent, needed another man. Experienced miners with good records who had quit to take some time off, holidays being unknown at that time, were usually taken back in short order but it was often a long wait for others. Bill had an advantage of sorts; an uncle in the investment business had introduced him to Chenoweth but, for the moment at least, there was no sign that the latter remembered him.

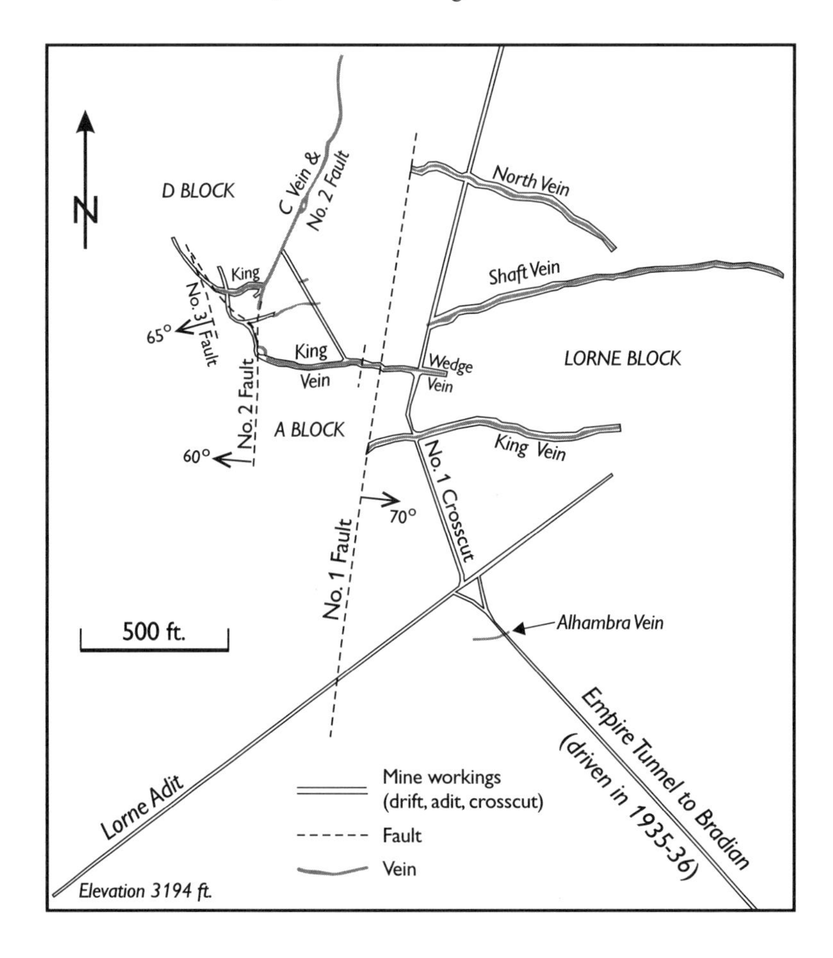


Plan of Bralorne's 8 Level in the King Mine block. Workings east of No. 1 Fault (except for the Empire Tunnel) were completed by Lorne Gold in 1929. They came so close, missing what would become the King Mine by less than 100 feet. The discoveries leading to it began in 1932 when Bralorne extended the drift following the Wedge Vein west through the No. 1 Fault to hit the faulted, offset continuation of the King Vein. After Hedley, 1935.

On arriving at Bralorne, Dunn and two friends had taken possession of a cardboard-covered shack across Cadwallader Creek from the mine. The former occupant, a bootlegger and moonshiner, with his still destroyed and now serving two months in jail, was unlikely be needing the place in the near future.

The trio spent most of their time waiting in the Bralorne line, although they made the occasional trip to Minto and Pioneer mines. The latter seemed to prefer Swedes and Finns and Dunn gave up there after Ed Emmons, the mine superintendent, looked him up and down and commented, "We don't need you here - too inexperienced."

It took about a month, but finally Chenoweth picked out Dunn from the Bralorne line. The next few days would be the acid test. If he was unable to handle the work, another would take his place. Dunn, never having worked underground and with only a vague idea of what was expected of him, was determined to make a go of it, especially after a hotshot young mucker grumbled, "If they put him with me I'll work him to death." He was, and the pair was assigned to a tough old Scot miner driving a crosscut on 11 Level of the King Mine.

Dunn's first day underground was the worst; on reaching the King Shaft four men at a time squeezed into a tiny skip* and were lowered down the inclined shaft from 8 to 11 Level. At the working face,* Dunn was motioned to the left side where, working cross-handed, he was expected to match his nemesis shovelful by shovelful. That day they mucked 19 tons of broken waste rock into one-ton ore cars and hand-trammed the loaded cars some 700 feet to the shaft. Worst of all, Dunn was sick to his stomach, a reaction to breathing powder fumes still hanging in the air of the poorly ventilated crosscut.

After three shifts, just as Dunn was getting used to the work and the powder fumes, the miners went on strike for higher wages. It left him in a bad spot. He had already run up a big bill at the company store, most of it for food for himself and his two friends, the latter still without work. The strike lasted 14 days, marked by some very slim meals, but when it was over the trio all had jobs, with their pay as muckers now $4.50 a day, an increase of 50¢. In the money at last, they left behind the bootlegger's shack and a dispute over their use of a nearby outhouse and purchased "Alcatraz Island" in Cadwallader Creek for $100. The latter had a bridge leading to it, a cabin and its own self-cleaning outhouse perched precariously over the creek. By the end of the first summer Dunn had managed to save $300, almost enough to cover another year at university. Aside from a break during the war years he stuck with it, coming back each summer before graduating as a mining engineer and going on staff at Bralorne, where he remained until 1947.

## Buddies and Blacklists

Experienced miners and engineers usually had an easier time of it, as most were hired through the "buddy" system. Mining in British Columbia had come into its own about the turn of the century, and a core of experienced men had stuck with the game through the booms and busts that followed. Many were Cornishmen and, as one put it, "Us Cousin Jacks has got to stick together."[2] Scandinavians formed another, equally clannish group. Managers and foremen were always on the lookout for good men they had known in other camps or who came highly recommended by trusted friends. On turning down an offer from a rival mine, a man might well suggest someone else for the job. Thus, though Bralorne and Pioneer were bitter rivals at hockey and their senior staff members stand-offish at technical meetings, information on potential employees was traded at intermediate levels. It cut both ways; a blacklist meant just that! Anyone suspected of being a slacker or, worse still, interested in unions had little chance of landing a job.

One victim of the blacklist was a young miner who quit Pioneer and went home to Winnipeg for Christmas 1937. Returning after the holiday, he expected to be taken on within a few days but instead waited in the line for close to a month before realizing that neither mine had any intention of hiring him. That was it; no explanations were offered and there was no opportunity for either a hearing or a protest. Thinking back, he assumed that he was being punished for sounding off to the Pioneer brass when they visited him at the working face deep in the mine. It happened on a bad day: the heat was oppressive due to poor ventilation, he was feeling ill and, to top it all, his helper was next to useless, considering hard work underground beneath his status as a Pioneer hockey player. Asked how things were going the miner snorted that he needed a good helper willing to work rather than his hockey player or perhaps a musician from the Pioneer band! There was no rebuke at the time and the miner assumed that the shift boss knew he was doing his best.

By now the young miner was desperate, flat broke with no hope of a job and unable to leave the Bridge River. Everything of value, including his watch, had already been pawned. First, there was help from a friend working at Bralorne who slipped him into the bunkhouse, letting him sleep in the bed of someone on night shift and smuggling him food, mostly pie and cake, from the cookhouse. He had to clear out during the day, hiding out of sight in the compressor or powerhouses, both comfortably warm.

The next offer of help came from an unexpected source, Zada Fontaine, madam of the "sporting house" at Ogden. They met when he was lounging in the Bralorne drugstore and Zada, a smartly-dressed woman of about 40, came in to pass the time of day until her girls returned from their routine visit to the Bralorne doctor. Sensing something troubling him, she asked what it was and, on hearing his story, offered him a place to stay, meals and a little spending money in return for looking after the wood and water supplies for her establishment. Her offer, much appreciated, marked the turning point in his fortunes. After a few weeks at Ogden he landed a job in the Bralorne store where, for some reason, the blacklist did not seem to apply. Then not long after, his transgressions either forgiven or forgotten, he was allowed to resume his mining career, with the only after-effect being an aversion to pie and cake lasting another 20 years.

## *Help for Job Seekers*

In an attempt to discourage large numbers of unemployed men flooding into the Bridge River the two mines combined in about 1935 to open a hiring office at Shalalth.[3] Anyone seeking work with them had to register there and, after doing so, some men waited nearby in crude shanties while others made the trip over Mission Mountain to visit the mines. The latter, if lucky enough to be offered something, might well be ordered back to Shalalth to re-register before starting work.

Bridge River's residents did their best to help the unemployed. Will Haylmore, the mining recorder, made an outstanding contribution by providing work around his office and home at Haylmore near the mouth of the Hurley River. In the words of one observer, "Any poor bum travelling through the Valley could always get a day's work from Will who used the labour to terrace his hillside holdings."[4] His help earned their pay; much of the work involved collecting large boulders from among the river gravels and piling them in dry walls used to hold back earth-filled terraces. Others worked on his placer lease nearby.

At Gold Bridge, housewives, no longer surprised when their husbands showed up with unexpected guests at mealtime, became expert at turning out impromptu meals from food on hand. Even the mines tried to help; at Pioneer, mine foreman Bob Eklof would often ask "Big Jim," the Chinese cook, if there was any leftover food for the disappointed job-seekers waiting in the line. With luck, that could mean a meal and sometimes sandwiches to eat on the ten-mile return hike to campsites near Haylmore. A full belly wasn't too much consolation for a man desperate for work, but at least he knew that someone else cared.

## On the Job

Mining the gold-bearing quartz veins of the Bridge River valley involved three steps: (i) sinking shafts to reach the veins at depth, (ii) driving levels or lateral workings, crosscuts to reach the veins, and drifts along them to sample gold values, and (iii) stoping, where blocks of ore between levels were broken, passed to the level below through chutes,* trammed to the shaft, and hoisted to surface for milling. Each mine had its own way of doing things, which changed through the years as new engineers arrived, equipment improved, and increasing depth brought new problems in the form of higher temperatures and rock pressures. Sinking shafts and driving levels was usually done by special crews paid on a contract basis, either a set price per foot, or wages plus a bonus for any footage above an agreed minimum. Highly skilled miners could earn far more than the average wage and, for a mucker, it was considered an honour to be asked to join a special crew and share in the bonus.

Operating the mines was a complex business involving many others besides the miners and their helpers. For a start, the pipefitters supplied each working with compressed air and water for the drill, and the ventilation crew provided fresh air for the miners to breathe and to clear out the powder fumes after each round was blasted. Timbermen took care of some of the more important timbering, and train crews handled the ore on main haulage levels. In addition, there were hoistmen, skiptenders, trackmen, electricians, surveyors, samplers, geologists and, of course, the supervisory staff. The name of the game was to keep the mill supplied with the tonnage and grade of ore desired and, while doing this, ensure the long-term future of the mine and hold operating costs within acceptable limits. The mines, now large and complex, had outgrown the earlier approach of just bulling ahead and trusting to luck.

Preparing to go underground the shift crew changed to their "diggers" or workclothes in the "dry,*" a large room usually equipped with lockers, showers and an overhead rope and pulley system for hanging clothes. Most "diggers" consisted of "bone-dry" pants and jacket of stiff canvas worn over heavy woolen underwear and, as footgear, heavy, steel-toed boots.

For illumination underground the men used the open flame of a carbide lamp* clipped into a bracket on either a soft canvas cap or a hardhat.

Advertisement for carbide lamps, *The Miner,* April 1937.

The latter, initially known as "hard boiled hats," were a relatively new development, made compulsory at Pioneer in 1934. The lamps, with chambers for water and carbide, gave off flammable acetylene gas when water, controlled by a simple valve, dripped onto the carbide, the latter when fresh resembling pea-sized chunks of crushed brownish rock. Lighted by means of a flint and wheel, similar to a cigarette lighter, the gas burned until all the carbide had reacted, generally about 45 minutes. Then the base of the lamp had to be unscrewed, the now-inert damp, whitish residue knocked out, fresh carbide put in, the water chamber topped up, and the lamp relit. Usually recharging was done in the light from someone else's lamp but there were times with no one else around when it had to be done in total darkness.

The company made carbide and water available on the working levels but each man was expected to provide his own carbide lamp. In addition, most carried spare carbide in a small metal can, shaped like a hip flask to fit comfortably in a back pocket, plus a knife and waterproof matchbox.

By late 1936, carbide lamps were replaced by electric lamps at both mines. Powered by a wet battery worn on a special, heavy leather belt, they gave a good light and could last up to 12 hours between chargings. Initially, some diehards objected to the change, unhappy over the weight of the battery or the general inconvenience. The new lamps were company property, picked up at the lamp room on the way underground and returned at the end of the shift for recharging. Each man had to provide his own miner's belt.

Heading underground, the men either walked or, if the distance was long, rode a train to the shaft they were to descend. At the shaft some eight to ten men squeezed into the cage.* After closing the doors and making certain that everyone was ready to go, the skiptender, who rode with them, signalled the desired level by ringing a sequence of bells that sounded both on the level itself and in the hoistroom above. Following confirming bells from the hoistroom and acknowledgment by the skiptender, the cage was lowered to the desired level. Communicating in this way, the skiptender ordered repeated trips until everyone had been delivered to the proper level. This done, the shaft was freed for transporting ore and waste rock, either in a skip suspended under the cage or by running loaded mine cars directly into the cage.

At the working face, rock or ore was broken by drilling a pattern of holes, loading them with dynamite, and exploding the charge by means of a slow-burning fuse leading to a detonator.

The drills, heavy machines operated by compressed air, were of two types: drifters* used for horizontal and slightly inclined holes and stopers* for vertical and steeply inclined holes. Drifters, weighing about 150 pounds, were clamped to a steel bar wedged in either a horizontal or vertical position, and the drill advanced by a feed screw. Stopers, about 90 pounds, used a telescoping tube forced out by air pressure to jam them in place against the underlying floor. In both, the holes were started with a short piece of drill steel,* 1⅛ inches in diameter, forged at the tip into a cutting bit 2¼ inches in diameter. As the hole advanced, longer drill steel was needed at 18-inch intervals and, at each such change, the diameter of the drill bit was reduced by ¼ inch to lessen the possibility of jamming. All

Drifting on the Pioneer Vein on Nine Level East, 1934. Leonard Frank photo 17370, courtesy of B.C. and Yukon Chamber of Mines.

the drill steel was hollow, and water forced through it helped to reduce the amount of dust going into the air.

After the drill pattern was completed, the holes were loaded with dynamite. The paper-wrapped sticks were split with a knife before being tamped into place with a wooden stick. In each hole one stick of dynamite was pierced with a wooden or copper pin, a detonator crimped on the end of a length of fuse cord inserted, and the fuse tied around the stick to prevent the detonator slipping out. Fuse burned at the rate of one foot every 40 seconds, and the sequence of explosions could be controlled by the length of fuse used and the order in which the fuse cords were lit or "spitted."

Blasting was done towards the end of the shift. Muckers and others retreated a safe distance, making certain that someone was stationed at all possible approaches to the blast area, and were joined by the miner once the fuses were spitted. Usually it was possible to count the explosions, and the crew would know if one or more shots misfired.

Eight hours after disappearing into the mine portal, the men reemerged, their day over. At Bralorne there was a climb of 127 steps up from the portal to reach the dry located on the bench above. By now, grey with dust and fatigue, the men all looked so alike that, in the words of one miner's wife, it was next to impossible to pick out one's husband among them. Some of the older miners, their lungs partially destroyed by silicosis* and too weak to tackle the steps, would shuffle along the gentler grade of the switchback road and, on reaching the dry, collapse onto a bench gasping for air. After showering, the grime was gone, but traces of the smell, a mixture of powder fumes and the garlic-like scent of crushed arsenopyrite,* clung to their bodies.

## *The Hazards: Accidents*

By their very nature, mines are dangerous places to work. Mechanical equipment and explosives are used in confined spaces where the enclosing rock, already shattered by explosives, may collapse without warning. Another unseen hazard lurked in the mine air: tiny dust particles that, over time, could destroy a miner's lungs. Fortunately, by the time Bralorne and Pioneer were hitting their stride the dust problem had been recognized and the first steps taken to reduce the danger.

The government of British Columbia has overseen the mining industry since the passage of the Inspection of Metalliferous Mines Act in 1897. Inspectors visit the mines regularly and, in the event of an accident, conduct an investigation into the causes. On observing unsafe practices, an inspector can order changes made within a reasonable time or, if the matter is serious enough, shut down the operation until such changes are made.

Some accidents are perhaps unavoidable but, with care and good co-operation between employees and management, the number can be held down. From 1933 to 1942 inclusive, 14 men lost their lives in mine accidents at Bralorne and Pioneer, a fatality rate close to that for all mines in the province.[5] Some of the accidents involved a breach of regulations, either through ignorance or, more likely, through cutting corners to save time and effort. The bonus system with its emphasis on production may well have been a contributing factor in some.

How many men does it take to clear a hung-up chute? Possibly a posed photo at Bralorne with timbermen, battery locomotive, motorman and helper, the latter (on the left) with a coil of fuse wire and detonator over his shoulder, 1934. Leonard Frank photo 17473, courtesy of B.C. and Yukon Chamber of Mines.

Rockfalls* are a constant hazard underground. Most occur in freshly-blasted workings but, less commonly, a section of the back or walls of an older working may drop without warning. At the working face, before mucking the round blasted on the previous shift, either the miner or a mucker under his supervision tested the ground with a steel pry bar. Any rock that looked ready to come down was pounded or levered in an attempt to work it loose. A dull, hollow sound on striking a solid-looking section might mean a hidden fracture and an air space between it and solid rock above. Careful barring down* and timbering reduced the danger but it could never be eliminated. Six of the 14 fatalities were due to rockfalls, several of them involving men regarded as particularly careful miners.

Chutes, used to feed ore broken in the stopes into mine cars on the level below, were another hazard. Filled with fragments of all sizes, they had a tendency to hang up or jam. Some could be freed by poking at them with a steel pry bar but others had to be shaken loose or "bulldozed" using a small piece of dynamite tied to a stick, pushed up into the chute, and detonated. Men working on the floor of broken ore in an overlying stope were expected to keep well clear of any hung-up chute for fear of being drawn in if it let go. Despite all the rules and, in two instances, specific warnings

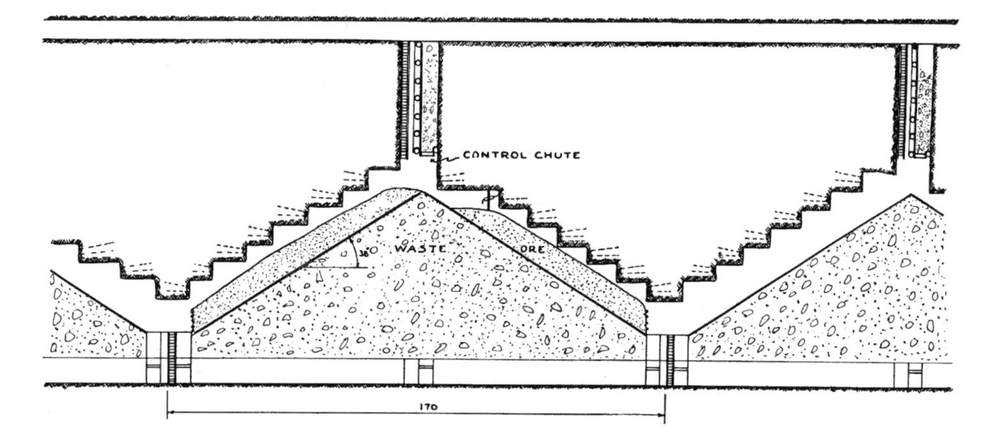


Sketch of a rill cut-and-fill stope. Planks are laid on the waste material and a height of about 7 feet is maintained at the working face. Other stope types include: cut-and-fill, where the ore is moved to the chutes by hand or by air-powered scrapers over planking laid on a horizontal floor of waste; shrinkage stopes, where sufficient ore is drawn down to provide a 7-foot working space between the top of the broken ore and the back, and on reaching the level above, the stope is drawn down, or emptied, before back-filling with waste rock from the top. *The Miner*, April 1937.

about the chute that trapped them, three men lost their lives in this manner.

Three others died in shaft accidents. In two instances, men on shaft-sinking crews were struck by falling objects while working in the as-yet-untimbered portion of a shaft where a bucket was being used for hoisting. In the other accident, a cage-tender, perhaps on hearing an unusual noise, apparently put his head outside the ascending cage, struck the shaft timbers, and was dragged out of the cage, falling to the bottom of the shaft. Another man died when he stepped or fell from an ore train into an open raise, inexplicably left with its wooden cover removed, and fell some 70 feet.

Only one of the fourteen fatal accidents involved explosives. A miner bulldozing oversize ore hung up on the grizzly,* or grating, where the ore was fed into the coarse ore bin of the Bralorne mill, spitted one charge but, on having trouble spitting a second, told his helper he would get it later. Putting the second stick of dynamite in his back pocket he walked back to the portal where he stopped an ore train and was talking to the crew when the shot went off in the grizzly house. He just got out the words, "All right boys, only one shot," when the second went off in his pocket. Horribly mutilated, he died a few hours later.[6]

Concerns for the safety of miners riding in the cages led to detailed regulations covering hoisting operations. The steel hoist ropes were inspected both routinely and after an accident with any damaged or defective portions replaced before the shaft went back into operation. Mine cages, travelling at up to 1,000 feet a minute, were held in position by wooden guides anchored in the shaft timbers. In the event of a hoist rope going slack, safety dogs, mounted on the cage, released automatically and pressed into the wooden guides to stop the cage. Hoisting of men and ore at the same time was prohibited. There were strict controls on the transport of drill steel and timber, the risk being that these items, if improperly secured in the cage, could work loose and snag in the shaft timbers.

Runaways in a shaft, either through a broken hoist rope or failure of the hoisting machinery, were rare. Pioneer had an unusual one in their No. 2 Shaft on Boxing Day, 1938. After checking the hoist machinery following the holiday break, the hoistman prepared to test run the two cages up and down the shaft, as required by regulation prior to transporting men. Switching on the electrical power and compressed air, he began to hoist the east side cage, but it had only moved a few feet before the drum carrying the hoist rope for the west side cage started to unwind at increasing speed. Safety controls and air brakes on the hoist mechanism failed to halt the runaway, and the cage hit the bottom of the shaft at 26 Level, some 3,250 feet below. The hoist rope, torn from the drum, followed and was a total loss. Assuming that moisture from the compressed air had frozen in the lines and control cylinder over the holiday, an attempt was made to thaw the lines. This ended when the east side cage began to run away, but fortunately the hoistman, reaching the controls in the nick of time, was able to stop the cage just as it reached the bottom of the shaft. On further investigation it was found that the compressed air line, also used to operate an air-pressure tank that forced water to houses on a bench above, had inadvertently been filled with water when a pumpman, not realizing that the main air valve had been shut off over the holiday, had tried to start up the water system. Shortly after, when the hoistman opened the valve, water had been forced into the hoist control mechanism where it had frozen, causing the failure. The cause determined, the compressed air line was changed at once to give a direct, individual air supply to the hoist controls.

Justifiably, the government's published report on the incident ended on a self-congratulatory note:

> There was no person injured in this occurrence, but it demonstrates the value of General Rule 96 . . . which requires: "After any stoppage of the hoisting machinery for repairs, and after a stoppage for any other purpose which exceeds four hours' duration, no person shall be raised or lowered thereby until the cage or skip has made one complete trip up and down the working portion of the shaft."[7]

## *The Hazards: Silicosis*

This debilitating and in some cases fatal disease with symptoms much like those of tuberculosis results from breathing mine air laden with microscopic particles of rock dust. Progress of the disease varied: some miners "dusted" years earlier would show only the primary stages, invisible except in chest X-rays, while others, particularly younger men, might die within a year or two of starting work underground. Much of the danger could be eliminated by cleaning up the mine air.

By the time the Bridge River began, the cause of silicosis had been recognized, and in 1933 the government took the first preventive step and banned dry drilling. In future, all drills were to be hooked up to water as well as air lines, and drill steel with a hollow core was used to carry water to the face of the drill bit. About the same time, water sprays came into use to settle the dust around muck piles and chutes.

Late in 1934, provincial Department of Mines personnel began measuring the dust content of mine air using a Zeiss Konimeter, one of a number of instruments designed for this purpose. In operation, a tiny sample of air was forced against a special glass plate to which the dust particles adhered. The diameter and the number of particles were determined by subsequent microscopic examination. Early results showed dust counts of about 6 to 15 million particles per cubic foot of air in samples taken close to a working face immediately after drilling and blasting. Some counts were even higher but, more disturbing, the particles, all less than ten microns (1⁄100 millimetre), showed little tendency to settle out with time, resulting in unacceptably high dust levels as much as nine hours after blasting.[8] Water sprays were of some help, but the obvious solution to the problem was to flush out contaminated air with a stream of fresh air under pressure. By 1938, all the larger mines in the province had trained engineers using the Konimeter to keep a careful check on dust conditions.

At the time dust counts were beginning, both Bralorne and Pioneer mines depended on "natural" ventilation, a system that utilized differences in elevation between mine entries to create an airflow. At Bralorne, cold winter air entered the workings through the main portal on 8 Level, at the time the lowest entry into the mine, and was heated in the workings to about 60°F and then, lighter than the outside air, rose upward escaping through the overlying entries into the King and Empire sections of the mine. The flow through the mine was directed and controlled by means of carefully-fitted, virtually airtight doors, brattices* of wood and canvas sealing off disused workings, and specially-driven ventilation raises. In hot summer weather, the flow reversed, with air cooled in the workings escaping out the 8 Level portal. "Natural" ventilation was adequate most of the time but rapid changes in temperature or barometric pressure could temporarily halt or reverse the flow of air. The next improvement was the use of booster fans and galvanized steel ventilation pipes, about 15 inches in diameter, to carry air into workings where ventilation was inadequate. In later years as the mines went deeper, natural ventilation was abandoned in favour of forced ventilation employing powerful fans to force down outside air and to draw spent air from the mine workings. In 1941, Bralorne installed what would prove to be the first of a number of systems, and Pioneer started up a powerful new system in January 1954.

Beginning in 1936, government regulations required that all employees working underground have a yearly medical examination and hold a certificate declaring them free from respiratory disease, The improvements paid off and silicosis does not appear to have been a serious problem among those who began their mining careers in the Bridge River in the 1930s. In contrast, some older staff members who had been in the mining game much longer were found to be silicotic and may well have had the disease for years without progressing past the primary stage. A number of these men, although not banned from underground until as much as 20 years later, lived to complete their careers and enjoy many years of retirement, all without any obvious ill effects.

The year 1936 also saw the passage of legislation granting compensation to the victims of silicosis. Benefits, initially a minimum of $10 a week for total disability, improved over the next few years. Lobbyists for the legislation included a number of miners' organizations and newspaper man George Murray, the MLA representing Bridge River in the provincial legislature. The mining companies went along reluctantly, unhappy about costs charged to them and convinced that many of the victims had contracted the disease years earlier when working in other provinces. But, on balance, they may have considered it worth the price in order to deprive the union movement of a hot issue in the growing campaign to organize the industry.

The Bralorne Hospital, probably as it was early in 1935. Photo by W.A. Hutchings.

# *Company Towns*

Bralorne and Pioneer townsites were developed by the mining companies, which had absolute control through ownership of both surface and mineral rights on their Crown-granted mineral claims. They did the planning; supplied services such as roads, water and sewer; built single men's accommodation and houses and, when all was ready, decided who could live there and under what terms. Company stores, medical facilities and schools were provided and, in addition, there was generous help in the construction and operation of recreational facilities. The latter, something of an exception, were operated by community clubs that had elected boards of trustees and were financed by a monthly levy on all employees.

A few outside businesses such as banks and utilities were allowed to

The office area of Bralorne townsite, 1934. Photo by Leonard Frank, Vancouver Public Library VPL 14849.

build on company land. Others had to find sites elsewhere, preferably as close as possible to their potential customers at the mines. Beer parlours and the sporting houses were two of the banned enterprises. It was not that the mining companies had anything against them; indeed it was more the opposite, with many officials considering them essential if they were to hold on to some of their better miners. It was just simpler to have them down the road on someone else's property. In contrast, gambling was more or less condoned, with both mines providing card rooms. At Bralorne prior to the opening of the Boultbee Memorial Church in mid-1936, Sunday services held in the theatre hall had to compete as best they could with the racket from the poker games downstairs.[1]

Initially, accommodation at the mines was in bunkhouses, with a few tiny houses provided for key staff members. As ore reserves grew and it became obvious that the mines had a long-term future, more and more houses were added and made available to the better miners, hockey players and others that the companies were anxious to keep on their payrolls. The companies maintained the houses and, by doing so, got involved in the inevitable bickering and complaining that starts as soon as one's neighbour appears to get preferential treatment.

At Pioneer, construction dated back to 1927, when log cabins were built on what would later become the school townsite. The main building pro-

The bunkhouses at Pioneer, looking southeast (upstream), about 1935. Photo by W.A. Hutchings.

The first Bralorne schoolhouse, 1935. Photo by W.A. Hutchings.

gram ended late in 1937 but, with the addition of a few more houses during and after World War II, the company had a total of 114 residences in 1954.[2] These were in sites spread out for more than a mile along the bottom of Cadwallader Creek's V-shaped valley, as there was simply no space suitable for a single, large townsite. Some of the houses were perched on sidehills and others on narrow flats in the creek bottom. Many were alarmingly close to the creek, so close that there were clotheslines strung across the creek, the only open space available. One surface foreman had to grin and bear it when his carelessly strung line collapsed into the creek and the family washing was swept away. There was another, more serious disadvantage to the location; the steep ridge to the south hid the sun's rays from early November until early February, turning it into a cold, grey pocket.

Bralorne had a much better spot for their first townsite, on a bench about 200 feet above the creek and open enough to the southwest to have sun on the shortest day of the year. There was room for company offices, bunkhouses, community hall, gymnasium, store, bank and a number of company houses. Beginning in 1934, before Bradian was recombined with Bralorne, the No. 2 Townsite was developed on a similar bench about half a mile to the southwest. Eventually it would hold about 70 houses, school buildings and a community church, plus ball park, tennis court and hockey rink.[3] Much later, beginning in 1947, a new building program added 50 houses in the Bradian Townsite a half-mile upstream of No. 2.

The company houses must have seemed an unattainable luxury to employees too junior to rate them or to business people struggling to build something at Gold Bridge or along the road allowance. Some less fortunate

The Bank of Montreal at Bralorne, 1934. W.A. Hutchings, manager (left), Geough Beley, clerk (right). Photo from W.A. Hutchings album.

mine employees squatted, living in log cabins or shacks built on land adjacent to the company holdings, where surface rights were held by others or retained by the government. The alternative was to move into a bunkhouse and, for the time being, give up hope of bringing one's family to the Bridge River.

In truth, the houses were simple structures, hurriedly slapped together using light frame construction and costing less than $2,000 each. Generally, they were built on posts resting on small cement piers, the basements having a dirt floor aside from a cement pad for a wood-burning hot air furnace. Floor plans for the living area varied, but in all the obvious intent was to cram as much as possible into a minimum amount of space. This led to some strange compromises, such as Bralorne's windowless or "blind" bedrooms, the centre one of three built against the common wall of a duplex. Another design cut down on hall space by having both front and master bedroom doors open directly into the living room. When Bralorne's Bosustow asked a foreman's wife how she liked her house she summed it up neatly, "Kitchen too small. Head in oven, arse out the back door."

In March 1937, Pioneer tried to improve things by involving their employees in house planning. Dr. James, managing director, offered prizes for the best four- and five-room designs, using a basic kitchen and living room plus either two or three bedrooms. There was no time to waste; entries had to be submitted within a few weeks and the houses were to go up that summer.[4]

Tenants were expected to supply their own firewood and, in a cold winter, it might take as much as ten cords of wood for heating and cooking in an average house. None of the houses were insulated. Wood came close to being the universal fuel in the Bridge River; at Bralorne even the steam plant used to heat the main buildings was fired with 46-inch lengths of logs. Firewood could always be bought at the going price, usually close to $3 a cord, but many families preferred to cut their own. In late summer and early fall they would go out cutting on the hillsides above the mines and their cordwood, piled and tagged beside one of the many wood roads, would be picked up and delivered to them by company truck. Some of the cutting, poorly controlled, added to the snowslide hazard.

The interior of the Bralorne store, 1934. Photo by Leonard Frank, Vancouver Public Library VPL 14855.

Pioneer mine, June 1934. Photo by C.E. Cairnes, Geological Survey of Canada No. 77812.

# Developments at the Mines

Aside from occasional setbacks, both Pioneer and Bralorne mines prospered until late 1942 when operations were cut back due to wartime restrictions and labour shortages. Bralorne, the upstart, became the larger operation when its production passed Pioneer's in 1937. Both companies would explore for and attempt to develop new mines both in the Bridge River and farther afield in British Columbia and western United States, but nothing significant came of their efforts.

## *Pioneer*

The company's best fiscal year ended March 31, 1935, when 89,786 ounces of gold were produced from 130,545 tons of ore milled. In the following years gold production fell off somewhat as the Main Vein was followed to increasing depths.

In June 1932, Dave Sloan had hired Dr. Howard T. James as mine geologist and engineer to assist in Pioneer's transformation into a major mine. James, with a mining engineering degree from UBC, had later taken graduate work in geology at Wisconsin and Harvard, obtaining his doctorate at the latter. He then worked for three years at the copper mines at Anyox, B.C. and for a year at the provincial government's Department of Mines office in Prince Rupert. Finding the latter required being away from his family most of the year, both in the bush and in Victoria preparing reports, he resigned and moved to Vancouver with nothing in sight. At a mining meeting, friends of James, aware that Pioneer could use more technical help, had recommended him to Dave Sloan.

At Pioneer mine the "Doc," as he was known, was expected to handle just about everything. That included surveying, something he had never done, but before he left Vancouver another friend offered help. Borrowing a transit from the university, the pair practised until James was confident he could handle it.[1] Unlike Sloan, a natural promoter, James was a shy, somewhat austere man, perhaps more scientist than engineer, who sometimes had difficulty understanding and communicating with the men working for him.

Within a few months, James proved his worth by persuading Sloan to extend a drift on 7 Level that had been stopped 800 feet east of the shaft.

There, the Main Vein, cut off by a fault, had been picked up in a crosscut and drifted for a few rounds only prove too narrow and too low grade to be of interest. After 400 feet of new drifting the workings entered an ore shoot where for 300 feet the vein, now 2–5 feet wide, carried gold values from $20 to $100 a ton and, best of all, gave no indication of an eastern limit.[2] The find, quickly confirmed on other levels, made a substantial addition to Pioneer's ore reserves.

Following this, anything "Doc" James wanted to do was fine with Sloan. In 1934, James was named general superintendent, replacing H.J. Cain who had resigned to go with one of the smaller Bridge River operations. E.F. "Ed" Emmons, a university classmate of James, joined Pioneer as mine foreman in April 1933.

Good news was the order of the day at Pioneer in 1934. On January 31, after a year of uncertainty, President Roosevelt had fixed the price of gold

Pioneer staff, 1934. From left to right: R.O. Udall, assistant superintendent; Ross Thompson, accountant; Paul Schutz, mill superintendent; E.F. Emmons, mine superintendent; Howard T. James, general superintendent; H.K. McKenzie, master mechanic; H. Ashworth, electrical foreman; J. Olsen, surface foreman. Leonard Frank photo 17282, courtesy of the Jewish Historical Society, Vancouver.

at $35 US an ounce, up from $20.67, the standard since 1792. Then on October 1, 1934, Pioneer's quarterly dividend was increased again, this time from 15¢ to 20¢ a share. Minor changes to the mill had increased capacity to 400 tons per day. The No. 2 Shaft, now being sunk to 26 Level, had reached 19 Level and new ore added to reserves more than made up for the tonnage mined. There were new discoveries on the upper levels, once thought to be worked out. The most exciting was on 5 Level where new drifting to the east had opened up 550 feet of ore with the vein averaging three feet in width and carrying 3.9 ounces of gold per ton. On the surface, growing confidence in the future of the mine was reflected by 26 new houses, a new hospital, community hall, tennis courts and a skating rink.[3]

By 1935, Dave Sloan was leaving much of the day-to-day management of Pioneer to Howard James. Perhaps it was to be expected; in his younger days Sloan had moved around a great deal and looking after a stable, well-established mine may not have had much appeal, or perhaps years of working 16-hour days had finally caught up to him. He had not been well in the summer of 1934 and early in 1935 had gone to Hawaii for a complete rest. By early summer he was back, deeply tanned, feeling fit and ready to go again.

On July 30, 1935, Sloan and R.W. Brock, Dean of Applied Science at the University of British Columbia and a prominent geologist, set off from Vancouver by Boeing flying boat on an inspection trip to the Pioneer mine. Originally Victor Spencer had planned to escort Brock but, by a quirk of fate, Sloan had insisted on replacing him, saying he could do a better job of showing Brock around, and besides, Spencer would be busy at the family department store. They made a brief stop at Brock's summer home on Alta Lake to pick up Mrs. Brock and, after taxiing part way down the lake, the plane, with four aboard, swung around and took off to the north. It was airborne and appeared to be climbing steeply when the single engine faltered and, after sideslipping a bit, perhaps as the pilot fought to regain control, the craft nose-dived into rocky, cut-over timberland at the end of the lake. The pilot and Dean Brock appeared to have died instantly, but loggers working nearby were able to extricate Mrs. Brock and Sloan from the wreckage. Both were terribly injured and, accompanied by two doctors and two nurses who chanced to be staying at Rainbow Lodge nearby, they were put aboard the caboose of a southbound PGE freight train and rushed to Squamish. There they were transferred to Spencer's yacht, *Dearleap,* and landed at Horseshoe Bay at 12:30 a.m., some eight hours after the crash.[4] Mrs. Brock died in the ambulance a few minutes later and Sloan on August 4, 1935. It was the end of an era.

Howard James was named general manager of Pioneer a few days later. No major changes resulted, but James, resident at Pioneer and lacking Sloan's outside mining interests, was now the man on the spot, free to develop the mine as he saw fit. A small sign of changes to come saw a stockbroker, allowed to operate in the Pioneer camp, relocate to Ogden a few months later.

By the end of 1935, No. 2 Shaft had been sunk to 26 Level, some 3,313 feet below the surface. Crosscuts had been driven to the Main Vein from 15 to 22 levels, and a limited amount of drifting done on the vein from 15 to 17 levels.[5] Much more drifting would be needed to confirm it, but from all indications the ore continued to depth and Pioneer's future was assured.

## *Bralorne*

Late in 1933 there were reports that Bralorne planned to concentrate exploration on the King Mine and spin off much of the ground between it and Pioneer mine to a new subsidiary company. This came to pass at a special meeting held on January 4, 1934, when Bralorne stockholders approved a deal that saw 26 of Bralorne's 58 claims turned over to a new, 2 million share company, Bradian Mines Limited. The Bralorne company would receive almost 1.2 million Bradian shares for the claims and Bralorne stockholders would be issued rights to purchase Bradian shares at $1 each on the basis of two rights for each five Bralorne shares held.[6] This issue of 400,000 shares was to be underwritten by two companies: Bralco Limited,[7] a new holding company controlled by Austin Taylor (25 percent) and Newmont Mining Corporation, a major American mining company (75 percent). In addition, the two companies were to receive options in the same proportion on a further 200,000 shares at $1.25 each. Rights issued to Bralorne stockholders expired on March 12, 1934, and the options at $1.25 on August 1, 1935. Austin Taylor was president of Bradian Mines and all the Bralorne directors were on its board plus two additions, Ira Joralemon and F. Searls, Jr., the latter representing the Newmont interests.

For Joralemon, becoming a director of Bradian and, by 1935, of Bralorne, marked a change of role. Initially, he had been a consultant making infrequent trips to the mine, but now, more and more, he was taking charge of exploration and development of the property. Dick Bosustow, in turn, provided day-to-day management while Austin Taylor, as president, handled corporate affairs.

Bradian Mines appears to have been spun off because Austin Taylor had no intention of ploughing Bralorne's growing cash reserves into a new

At the Bradian camp, 1934. From left to right: Ira Joralemon, consulting geologist; Austin Taylor, president of Bralorne and Bradian mines; a geologist, possibly Percy Dobson, Neil McQueen, Director of Bralorne and Bradian mines. Photo coutesy of the Mining and Metallurgical Society of America.

exploration project certain to cost several hundred thousand dollars. Perhaps it was as well from the operating standpoint too. The Bradian crew, housed in a 50-man camp just over a mile from Bralorne, could be kept busy on exploration, free from nagging worries about meeting production schedules. Initial plans called for two shaft-sinking projects. The first, on the old Coronation workings, involved deepening a shaft already down 200 feet for another 500 feet, and the second, some 2,000 feet to the northwest, involved extending the lowest Ida May tunnel into firm ground and then sinking a shaft for 460 feet. This done, the intent was to join the two shafts by a crosscut running from the 200-foot level of the Coronation Shaft to the bottom of the new Ida May Shaft.

Bralorne's fourth and Bradian's first annual meetings for the year ended December 31, 1934, were held one after the other on April 29, 1935, in the Hotel Vancouver. There was much good news and Bralorne had joined the ranks of the dividend payers with a total of 62½¢ a share in 1934. Mill capacity had been increased to 400 tons per day and the entire camp rebuilt. Yet, despite the heavy outlay, the company had closed the year with about $300,000 in cash on hand and, in addition, $125,000 worth of bullion in transit.

Reporting on developments at the mine, Joralemon wrote that a remarkable orebody (120 feet long and six feet wide, averaging five ounces of gold per ton) found on 7 Level of the King Mine at the intersection of the King Vein with No. 2 Fault, had been responsible for the high earnings in the first few months of the year. Some of this rich ore still remained in place beneath the floor sill on 7 Level. Elsewhere, there had been other, less exciting discoveries, and ore reserves at year-end were put at 300,000 tons of uncertain grade but "certainly as much as 0.35 ounces per ton, and the scattered rich areas will probably make the mill heads* average considerably more than this."[8]

In answer to questions from the floor, Dick Bosustow, mine manager, stated that no further capital expenditures were contemplated aside from a few more houses for married employees, and that, despite rumours, the present shaft and equipment were adequate to handle the increased mill tonnage, and that no difficulty had been encountered in supplying adequate ventilation to the workings. And he intimated that the present dividend rate could be maintained, provided, of course, that additional reserves could be developed beyond the present two-year supply.[9]

Shareholders at the Bradian meeting had already received a report by Joralemon dated March 26, 1935, and included in that company's annual report. In it he stated that the two shafts had been sunk as planned and that from the 7 Level at the bottom of the Coronation Shaft a 180-foot crosscut had been driven to the Coronation Vein. The vein had been drifted in both directions for a total of 470 feet, but, aside from a 70-foot section averaging 1.25 ounces of gold per ton over a width of 1½ feet, the vein was too narrow and too lean to be of interest.

At the Ida May, exploration had been complicated by the Empire Fault, a major structure that displaced the Ida May Vein, and a crosscut driven from the bottom of the 450-foot shaft, expected to cut the vein at 50 feet had finally hit it at the 290-foot mark early in 1935. The limited amount of drifting done on the vein since had disclosed some ore which Joralemon considered very encouraging and he concluded his report, "In the many thousand feet of strong veins on the property there certainly should be several valuable orebodies."

Bosustow had some better, up-to-the-minute news for Bradian shareholders. On the Ida May Vein, drifting to the east had developed a 150-foot long oreshoot, with an average width of 3.6 feet and a value of 0.328 ounces of gold per ton, and then, following a lean zone, word had been received from the mine that morning that the face of the drift was once again in typical high-grade ore, showing much free gold in a fine-looking ribboned quartz vein. In the drifting to the west the Ida May Vein had been cut off at the 200-foot mark by the Empire Fault. But values had been picking up as the drift neared the fault, some of the samples running as high as one ounce of gold per ton, and it seemed reasonable to expect good values when the crosscut, now being driven, picked up the offset portion of the vein. At the moment there was no thought of installing a mill, but that could easily change in the next few months as there was no intention of putting Bradian ore through the Bralorne mill.[10]

The Bradian project was Joralemon's baby and, in the early stages at least, those on the job found it hard to understand his enthusiasm for it. The money was there; Bradian still had close to $300,000 in the bank at the end of 1934, plus $100,000 more if an increasingly reluctant Newmont could be persuaded to pick up the outstanding portion of its share option. On his visits to Bradian and Bralorne, Joralemon would review the work of the mine staff and then make his own inspection underground. He worked quickly; on examining a drift for the first time he would walk slowly, pacing off the distance and mapping as he went along, produce a credible map with all the essential information on it. Back in the office, Joralemon, often writing with a stub of a pencil, would dash off his report, letter perfect and ready for final typing. He understood the necessity for making decisions and his, usually right, were made quickly without undue agonizing. Not an arrogant man, but confident of his ability and the value of his work, he would sometimes quip about his fees, "Tap on rock $5; knowing where to tap $500."[11]

Joralemon's faith in the Bradian project paid off. In July 1935, with developments looking better day by day, the Bradian company was merged into a fast-fading Bralorne company. Within weeks work began on the Empire Tunnel, a new haulage more than a mile long, to connect the workings on Bralorne's 8 Level with the Ida May Shaft, now renamed the Empire Shaft. On completion, ore from the Bradian section would be hauled directly to the mill on 8 Level, meaning a minimum of hoisting and transferring as well as freedom from weather and road conditions on the surface. But that was all about a year away, much too long to wait, and less than a week after the takeover a local trucker had a contract to haul 50 tons a day of Bradian ore to the Bralorne mill.[12]

Work on the Empire Tunnel was rushed, with three six-man crews working 24 hours a day, seven days a week, advancing it an average of 18.3 feet a day. A seventh man assigned to each shift allowed everyone one day off each week. Much of the hand labour was eliminated by use of a mechanical mucking machine operated by compressed air and by a battery locomotive for tramming. The tunnel, 5,384 feet long, started from the Bralorne end on August 15, 1935, was connected to the Empire Shaft on June 6, 1936.

Before the Empire Tunnel reached the former Bradian workings in late April 1936, extensive work had been done at the Empire Mine, with important orebodies developed on the Empire (formerly Ida May) Vein on 6 and 10 levels, the latter reached by a crosscut driven from the Coronation Shaft. Subsequently the Empire Shaft was deepened from 6 to 10 Level, connecting with both the Empire Tunnel and the lower workings. West of the Empire Shaft, the Blackbird Vein, believed the continuation of the Empire Vein but displaced hundreds of feet north by faulting, was developed between 5 and 8 levels and, at the end of 1936, nearly a quarter of the known ore reserves were on the Blackbird.

With this connection almost all the ore now came from the Empire Mine, allowing for "more leisurely and careful mining of the . . . scattered ore in the King Mine, with resulting higher grade of ore from that mine."[13] "Leisurely" certainly but "careful" perhaps, as in its last years the King Mine would served as a haven for Bralorne's hockey players, hired to win

Bralorne staff, about 1934. From left to right: unknown; unknown; Fred Gray, mill superintendent; C. Stevens, storekeeper; E.J. (Ted) Chenoweth, mine foreman; R. Bosustow, manager; Don Matheson, engineer; Dr. D.M. King, medical doctor; T. Sturgess, warehouseman. Photo courtesy of Lois Bett.

Bradian mine camp, 1934. Photo by Leonard Frank, Vancouver Public Library VPL 14852.

games, not set production records. In effect, in a sort of reverse takeover, the Empire Mine, its reserves approaching the million-ton mark, had now become the Bralorne Mine.

By now, veins were being referred to by number rather than by name, the Empire (Ida May) becoming the 51 Vein and the Blackbird, the 55. During 1936, two new, unexpected veins were discovered, the 53 Vein west of the Empire Fault and the 59 Vein, east of the fault but north of all the earlier workings. Both were short but between them the tonnage added almost equaled the total amount mined during the year.

In the latter half of 1937 a new underground shaft, the Crown, located about 1,300 feet northwest of the Empire shaft, was sunk from 8 to 14 Level. Plans were made to deepen both shafts to 20 Level, a project that was completed in 1940. There was still ore, lots of it, and, unlike the King Mine, no indication of bottoming. But as Joralemon wrote in Bralorne's 1939 report, finding it was no simple matter:

> The Empire-Blackbird vein system is exceedingly complex. In

> addition to the division of the vein into three main branches, towards the West and downward, in many places the quartz splits into two or more strands separated by a few feet of barren rock. Often the points where such strands leave the main fissure can hardly be noticed. Many short diamond drill* holes must be run to keep from missing valuable orebodies. This drilling is cheap compared with crosscutting and it is proving a most valuable aid to solving the many structural problems. Since the beginning of 1940 it has found rich orebodies on 3 and 4 Levels as well as on deeper levels.[14]

During 1940, ore development on the new 20 Level was even more successful than anticipated. The veins worked on the levels above were there undiminished in size and grade, plus there was a new find, the Coronation (77) Vein, lying unsuspected in the footwall* of the Empire (51) Vein. Over time the new vein, believed to be the blind,* westward continuation of the vein explored in the old Coronation property, would prove Bralorne's most productive, with orebodies from 14 to 45 levels.

Richard (Dick) Bosustow, Bralorne's General Manager and by 1935 Managing Director, died May 17, 1940, following a prolonged illness. An excerpt from a directors meeting held soon after reads in part:

> His outstanding ability and service were largely responsible for the Mine becoming the successful operation that it is at the present time. Any reference to Dick Bosustow would be insufficient without mention of what was probably his greatest attribute, that is his great capacity for understanding which endeared him to all who were privileged to know him.[15]

Don Matheson, a mining engineer who had been with Bralorne since 1931, now took over as general manager and E.J. (Ted) Chenoweth moved up to general superintendent.

## *The Yalakom Rush*

The Bridge River came down with gold fever again in midsummer 1941 on news of an exciting discovery made some 16 miles northeast of Minto in the Yalakom area of the Shulaps Range. Actually, the find dated back to the summer before, when Thomas Illidge and William White staked the two key claims, Elizabeth 1 and 2. The news got out when Illidge returned to stake two additional claims accompanied by friends of his, John Soppit and Sid Wilson, who staked blocks of eight claims each.

Apparently, on the way into Haylmore to record their claims, Soppit, Wilson and possibly Illidge stopped in at Ben and Myrtle Cromer's cabin at the head of Liza Lake. The Cromers, who had moved to the remote lake from San Francisco, were much involved in prospecting, hunting and fishing in the area. One version of what happened next is that Mrs. Cromer asked for a sample of the high-grade gold ore on the pretext of mounting it in their fireplace and that, soon after, both the sample and Ben Cromer appeared in the Bralorne office.

Don Matheson reacted quickly; Bralorne had to have the property and there was no time to be wasted. Words alone could never have convinced him; Matheson, known to make crowing noises at the sight of free gold in a working face, must have made his decision while handling the sample of white quartz, somewhat rusty-streaked, that according to the *Bridge River-Lillooet News* carried an estimated $3,000 to $5,000 a ton in wire gold.*[16]

An outfit of sorts was thrown together and a party of about a dozen men led by Matheson started out the next day. Most were junior engineers and all they knew about the expedition was a warning given the afternoon before to be available for a little "jaunt" on Sunday. Nothing had been said about either destination or the length of time they might be away, and one unfortunate would end up working for two weeks without as much as a change of socks.

The trip in was difficult. From the main road in the Bridge River valley a good, seven-mile pack trail led to the Cromers' cabin at Liza Lake where the party spent the first night. From there a rough trail climbed 4,000 feet in six miles to a pass through the Shulaps Range leading to Blue Creek, a tributary of Yalakom River. A few lucky ones rode Cromer's horses but the rest walked, crossing the pass in a snowstorm. The best showing lay about 2½ miles beyond and was exposed at the edge of a retreating snowpatch on the north wall of Blue Creek at about 7,500-feet in elevation, some 500 feet below the ridge and more than 1,000 feet above the valley floor. Camp, such as it was, was set up on the hillside close to the showings.

If the showings looked promising, the first step would be to stake a large block of ground around them, hence the large party, because each man was entitled to stake an eight-claim block. On a quick examination, there were strong similarities to the Bralorne deposit. Ribboned quartz veins, in places almost four feet wide, cut a porphyritic quartz diorite not unlike some of the host rock at Bralorne and Pioneer mines. What differed was that this diorite,* less than a mile across, formed an island in a mass of serpentine* and related rocks that extended for miles in all directions, quite unlike the narrow band of serpentine found at the mines.

Impressed by what they saw, the Bralorne crew went to work. Over a hundred new claims were staked, freeing most of the party to return to Bralorne. Ralph Sinke, the young engineer left in charge, spent hours in the cook tent with Bill White, a partner of Illidge's, trying to make a deal for the original claims plus three blocks of eight claims staked by others, including Ben Cromer. The claim holders, working together, had agreed that none would sell without consent of the others. Moving fast, Bralorne had the jump on any potential opposition and, within a few weeks, held an option on the property and had staked close to 140 additional claims. Bralorne's higher-ups were caught up in the excitement too, both Austin Taylor and Ira Joralemon appeared at Bralorne, the latter examining the new find in late August 1941.

For the moment, supplies for the Bralorne party came in over the Liza Lake trail. Ben Cromer and others looked after the pack-strings while, at Bralorne, Charlie Cunningham expedited and did his best to make sure that supplies arrived in the order needed. "Curly" Evans and "Big Bill" Davidson had attempted to scout an alternate route to the property from the Bridge River but had been forced to conclude that the only feasible route was a road up the Yalakom River east of the Shulaps Range, servicing the camp from Lillooet. Already, Bralorne Mines was offering to spend $5,000 on such a road if the provincial government would come up with a matching grant. George Murray, local MLA and owner of the Lillooet paper, was lobbying hard for the Yalakom route, having already travelled it on horseback to visit to the Elizabeth property.

The government's response to the lobbying was to have Dr. M.S. Hedley, a geologist with their Department of Mines, examine and report on the property. Hedley was well qualified for the task; his background including a year as Bralorne's geologist in the mid-1930s. He reported that the veins on the Elizabeth property had been staked in 1934, when a large area was blanketed with claims that were later allowed to lapse with no work done on the ground. White and Illidge had met at the government's training camp for prospectors at Cowichan Lake in 1939 where the former, a graduate geologist, was an instructor and the latter was a trainee. Presumably Illidge, either aware of the earlier staking or on rediscovering the quartz veins, had asked White to accompany him on the 1940 trip when the first two Elizabeth claims were staked. Illidge was still working on the property, currently in charge of stripping operations for Bralorne. In describing the veins as exposed at the time of his visit, Hedley commented in part:

> No accurate idea as yet has been gained of the average value of the showings beyond the fact that certain known sections are undoubtedly of high-grade. Systematic sampling has not yet been done.
>
> This is a raw prospect that merits investigation, which it is understood the company is fully determined to give it.[17]

That wasn't good enough for George Murray and he let fly at mining engineers in general and government ones in particular:

> If [mining engineers] tell the truth and praise a property they may get into trouble. If they don't tell the truth and don't praise a property, they may also get into trouble. So they become expert writers of English prose. They visit a property, then they make a report which is made so colourless and dreary that few read it. They throw in some big words in order to mystify the common herd. In hidden ways they communicate to the intelligentsia what they really have in mind.
>
> Now if Hedley boosted [the Yalakom], the place would become infested with prospectors and that would not do for those who are eager to get the cream of the country under their own stakes. If he boosted the country and the big strike petered out, then the whole government would be blamed. In short if he expressed himself boldly and ably as he is quite capable of doing, you cannot tell what would happen. In fact it would be better perhaps to cease sending government engineers into places save to advise prospectors.[18]

Murray, upset, had overlooked one firm conclusion in Hedley's report: a mining operation, if it developed, would have to be serviced by the Yalakom route. Shortly after, the government, willing to gamble that the find might amount to something, authorized the expenditure of $10,000 for a new road. Work started soon after and, when shut down by weather at the first of November, the road had been pushed to within 14 miles of the property.

Bralorne's work on the property ended in mid-October when poor weather made further work at the 7,000-foot elevation out of the question. By then five diamond drill holes totalling 760 feet had been put down and stripping done over a distance of 1,750 feet along four veins. Three of the veins were subparallel and lay within a few hundred feet of one another, while the fourth, somewhat farther away, was nearly at right angles to the No. 1 vein. As exposed, only the upper end of No. l vein, close to the snow-patch, carried visible gold but, judging by experience at Bralorne, this might well change at depth. In addition to the work on the veins, a number of buildings had been put up at the campsite lower on the slope, the largest a log bunkhouse 11 by 18 feet intended to accommodate 12 men.[19]

The bonanza, if indeed there was one, remained hidden. If plans were made to rush the work in the 1942 season they were set aside after the Japanese attack of December 7, 1941, on Pearl Harbor. From then on exploration targets were strategic minerals, essential to the war effort. In a report dated March 19, 1942, Joralemon commented: "Development of gold prospects, including the Yalakom . . . has not been conclusive, and further extensive work must be postponed until after the war."[20]

Years later when that time came, the Yalakom would prove a will-o-the-wisp.

# *Stock Market Capers*

The Bridge River boom both created new jobs and gave "investors" a chance to try their luck in the resulting action on Vancouver and eastern stock exchanges. Some two-thirds of the companies involved were newcomers, incorporated in late 1932 or in 1933. Times were hard but there was still money out there if the pitch was right. Promoters, brokers and company insiders went all-out to push their favourites.

The mining properties varied. Some had untested gold-bearing quartz veins strikingly similar to those at Pioneer and Bralorne, while others, lacking any kind of showing, could only boast a location close to a producing mine or a hot prospect. Names were part of the game; none of the new companies tried to use "Bralorne" but there was a "Pioneer Extension," and perhaps a dozen others managed to work the words "Bridge River" into their titles.

Some who bought shares were believers, in for the long haul, but most, given the chance, were only too happy to sell out for a quick profit. One of the non-believers, a young university graduate employed as the secretary of two of the new mining companies, played the market but never held a stock overnight, let alone took delivery. When he saw a stock start to move as the "pools" moved in buying blocks at ever-increasing prices, he would buy too, watch for the moment of hesitation marking the top and unload fast. His scheme worked and in most months he was able to clear $50, a big help in supporting parents hard hit by the Depression.[1]

## *Pioneer*

"Stocks don't go up by themselves, each stock has its daddy."[2] The words were "Sell 'em Ben" Smith's, an iconoclastic New York stockbroker who, some time in 1931, signed on as Pioneer's daddy by acquiring a block of about 200,000 Pioneer shares. Using a combination of skillful promotion and buying with one hand and selling with the other, Smith, "an artist at hiding a trade,"[3] manipulated the price of Pioneer stock. Over the next few years the Pioneer caper was just a small part of his wide-ranging activities.

Gold was one of the few things Smith was bullish on, convinced its price would rise through devaluation of the United States dollar. In mid-August 1931, Smith and his wife had sailed north from Vancouver aboard

# MIX GOLD

## MINES LTD. N.P.L.

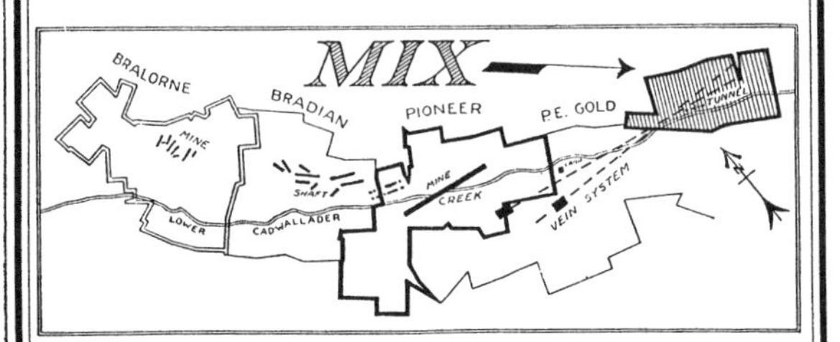


## *Situated 3500 Feet Upstream from Pioneer*

- Face of crosscut tunnel now in 840 feet.
- Last 34 feet in highly mineralized greenstone impregnated with quartz.
- This is the same formation and condition as occurs in Pioneer prior to cutting the veins.
- A fissure vein of Pioneer type expected to be cut at Mix within the next few feet.

---

**AUTHORIZED CAPITAL:**

**2,000,000 shares of 50c each. Issued in full payment for property and held in escrow under Government supervision—600,000 shares.**

**BOARD of DIRECTORS:**

| President, | Vice-President, |
|---|---|
| E. A. MARKHAM | ALEXANDER RUTHERFORD |
| H. D. CAMERON | J. A. GOODE |

# Northern Investors Ltd.

**Hall Bldg.** **TRINITY 6034** **Vancouver, B.C.**

This ad, which appeared in the *Financial News & Mercantile Review* for February 8, 1935, was typical of many that touted properties of real or imaginary potential in the Bridge River region.

the *Prince Henry* to inspect the Alaska Juneau company's gold mining operation in which he held 100,000 shares,[4] and on returning to Vancouver he appears to have flown to the Bridge River to check out Pioneer. One press report has it that, impressed by the potential of the mine and that no one seemed to be trying to unload the stock, he had offered to buy up to half a million shares if the owners cared to name a price. Pioneer's principals hesitated at first but then offers began to follow Smith on his travels. The first, for 25,000 shares reached him in Seattle, followed by one for 50,000 shares in San Francisco, both turned down as too small to be of interest, and later for 100,000 shares at $2.25 in Chicago and for another 100,000 at $2.70 in London, both of which he accepted.[5] Small amounts of Pioneer stock had traded in Vancouver as early as July 1928, but it was late 1931, just prior to listing on the Toronto and New York Curb exchanges, that the price hit $2.70, the cost of some of Smith's shares.

Smith's full name was Bernard E. Smith and he hailed from one of New York city's shabby Irish neighborhoods, where he had grown up about the turn of the century. A school dropout, in his time he had sold newspapers, simply bummed around, and held any number of minor jobs until a prominent stockbroker sponsored his membership in the New York Stock Exchange in 1926.[6] Later, in the aftermath of the 1929 crash, he acquired his nickname when he rushed into the crowded board room of the brokerage house where he had an office bellowing, "Sell 'em all! They're not worth anything!"[7] Thumbing his nose at the Wall Street establishment and their attempts to stabilize the market, Smith would go on to reap a fortune through massive short selling over the next few years. Put simply, short selling involves selling a stock one does not own in hopes that the price will fall before delivery is called for. Profits, or losses if the price rises unexpectedly, can be spectacular and Smith was reputed to have cleared $10 million in bear market operations in 1930.[8] A stocky, thick-necked man in his early forties, sometimes characterized as a "Wall Street pirate" Smith played the part with a brusque bantering manner and, at times, heightened the effect by donning a black eye patch to cover an eye lost in a boyhood accident.[9] He had a pirate's nerves too: whether it involved selling short a stock he considered overpriced despite a rising market or running a pool to manipulate the price of a stock.

In December 1932, spectacular high-grade was found at Pioneer in a stope above 8 Level, close to a serpentine body at the western edge of the ore zone. The high-grade, brought to the surface in wooden powder boxes, was amalgamated rather than being put through the cyanide mill. One 900-pound lot treated about this time yielded nearly 400 pounds of gold,

A member of a Vancouver Board of Trade on tour to the Bridge River area hefting a gold brick made great publicity for the mines. H.J. Cain, superintendent of Pioneer mine is second from the right. *Vancouver News Herald,* June 2, 1933, Public Archives of B.C. photo F-08984.

making it worth roughly $100 a pound at the gold price of $20.67 an ounce then in effect.[10] Sloan, in Vancouver for the holidays, made a rush trip back to the mine, returning with two samples, one of 24½ pounds worth an estimated $5,000 and a second of 72 pounds but not quite as rich.

Sloan's plans to display the samples had to be cancelled when Ben Smith, with a mining engineer and potential investors in tow, arrived in Vancouver and had to be escorted to the mine to see for themselves. On their return the smaller specimen was placed on display in a window of O.B. Allan's jewelry store on Vancouver's Granville Street. It consisted of pure, milk-white quartz laced with gold, the metal appearing to make up about a third of the volume. Ben Smith bought the larger sample for

$7,500 and took it to New York.[11] One story, probably apocryphal, has Smith showing it to New Yorkers and telling them that it came from a mine up in Canada where the locals think it's iron pyrites.* Whatever the pitch, the specimen was a remarkable advertisement for Pioneer and, within six months, Smith's activities together with other exciting developments would more than triple the price of Pioneer stock.

The stock, added to New York Curb and Toronto Stock Exchange listings in December 1931 and January 1932 respectively, traded actively in 1932 with a modest increase, bringing the price to almost $5 a share at year end. The largest volume of trading was in New York, followed by Toronto, with Vancouver trailing as a poor third. It was a perfect setup for Ben Smith, as the price could be influenced by playing one exchange against the other. In addition, there were fluctuations in the value of the U.S. dollar, at times worth close to $1.20 Canadian. Even weekly price swings, often 50¢ or more on the New York Curb, could be played: buying low and selling high, with the spread enough to cover brokerage and yield a profit as a short-term investment.

Trading volume in Pioneer stock picked up early in 1933 along with a gradual rise in price. Then, in a remarkable 15-week surge between mid-April and the end of July, over 1.9 million shares were traded on the three exchanges with the price rising from the $6 level to highs of $16.25 in Toronto and $15⅞ in New York. Initially, the activity may have been sparked by President Roosevelt's announcement of April 19, 1933, that the United States was abandoning the gold standard. Following this, there was much good news from Pioneer: on May 17, the release of a glowing annual report, a few days later word that the quarterly dividend rate would be raised from 6¢ to 15¢ a share, and on May 31 optimistic press reports on the annual meeting.

There were two wild price swings during the run-up, with most of the action on the New York market. The first came on Thursday, June 15, 1933, when Pioneer stock, which had hit $14¾ two days earlier, opened at $13½, fell to a low of $9⅞ during the day, then recovered somewhat to close at $11½ on a volume of 28,500 shares on the New York Curb. Gold stocks in general had been advancing for some time and when the correction came, the bottom may have dropped out of the market as speculators holding stocks on margin were sold out on stop-loss orders.

Ben Smith may or may not have had something to do with the first price swing, but the second, almost certainly his handiwork, was pulled off at the same time as Smith's caper in the stock of American Commercial Alcohol. Following the June 15 debacle, Pioneer stock had traded in the $12 to $14 range until Tuesday, July 18, 1933, when trades surged to 43,400 shares, more than most week's volume, with the price up $1¼ on the day to close at $15. Next day Pioneer peaked at $15⅞ on a volume of 46,900 shares, only to close at $13½ after hitting a low of $13. The slide continued for the rest of the week, and the stock closed at $11 after hitting a low of 10⅛. That was it; the party was over, the host had vanished and the celebrants were left nursing hangovers.

Pioneer's Sloan and Spencer may not have been aware that Smith, a loner, was managing the pool in American Commercial Alcohol stock under the very noses of a United States Senate committee investigating manipulation in the market. The stock was a natural; with the end of Prohibition in sight it was easy to convince would-be millionaires that the firm would soon be adding drinking liquor to their line of alcohol antifreeze. Buying and selling, Smith kept the stock in the public eye, working the price up from the $20 level at the beginning of May 1933 to a high of $89⅞ on July 18, 1933, the day he pulled the plug. Over the next three days the price plunged to under $30 and the pool account was closed. By the time the Senate committee realized what was going on and began looking for Smith, he was in Melbourne, Australia, a safe haven from their subpoenas.[12] Stricter regulation of the stock market was inevitable and the new watchdog, the Securities and Exchange Commission, began operating on the first of July 1934. With Pioneer traded on the New York Curb, company directors had to disclose their holdings, now just under 30 percent of the shares initially issued to them.[13] Smith, as Pioneer's "daddy" or market maker but not a director, did not, but his wings were clipped too, inasmuch as he would be the first to be questioned about any unusual trading patterns in Pioneer stock. Activity fell off and, with the big surges gone, the price of Pioneer shares held above the $10 level through 1934, then hovered near it until the summer of 1936 when a long, slow decline dropped the price to a measly $4.95 on the Toronto exchange shortly before Christmas 1936.

In 1933, with Pioneer looking like a sure winner, the three "Ss," Sloan, Spencer and Smith, combined to build the Gun Lake Lodge as their private retreat. Its actual location is on Lajoie Lake, or Little Gun Lake as it was then known, on land obtained from Mat Forster, a prospector who had held the ground for 25 years. During that time he had done stripping and a thousand feet of tunnelling on a narrow quartz vein that, in places, gave erratic high gold assays. He refused to give up and, in the words of the mining inspector, "If perseverance has any reward in mining, Mr. Forster should get it."[14] When Dave Sloan and Bob Eklof made the first trip to negotiate for the land they wanted, Forster refused to look directly at them and mumbled that they didn't have enough money to buy it. It was different on a second trip; a friendly Forster told them to go ahead and stake out the land they wanted.

The builder, Big Bill Davidson, a well-known Bridge River character, would have nothing to do with Ben Smith's suggestion that American carpenters be brought in. Using local men, he insisted that everything be done right, no matter how long it took, to "prove to those bastards we have the best carpenters in the country."[15] Davidson and crew built well. The lodge, a magnificent, two-storey structure built almost entirely of peeled logs, is still a showpiece that, until a few years ago, was in use as a lodge and could well be again in the future. In contrast the three "Ss" had only a few years to enjoy it. When it was sold in the fall of 1940, Sloan had been dead for five years and Smith, in hospital in New York, would never return to the Bridge River.

Between 1933 and 1938 Ben Smith made a number of brief visits to the Bridge River, most of them mere stopovers on his rush to somewhere else.

Little Gun Lake Lodge, built by Pioneer's three "Ss" (Sloan, Spencer and Smith), about 1934. Photo by Leonard Frank, Vancouver Public Library VPL 10102.

Undoubtedly, Dave Sloan and Colonel Spencer kept him up to date on developments at the mine, but there were times when Smith wanted to see for himself or to show the property off to would-be investors. Sometimes he travelled by train, leaving his private car on the siding at Shalalth and, on other trips, he would arrive by float plane, flying direct to Gun Lake from Vancouver or some distant point. Forever wheeling and dealing at a frantic pace, Smith seemed unable to slow down enough to appreciate the magnificent setting of the Gun Lake Lodge. Perhaps he came closest in the fall of 1934 when his wife and three of their children joined him for a brief holiday.

Pioneer's success generated much new wealth, and the three "Ss," Sloan, Spencer and Smith put some of it into other mining ventures, none particularly successful. Included were Wayside Consolidated Gold Mines, a Bridge River property long on lawsuits and scandal and short on ore, and B.C. Nickel Mines, a loser in an attempt to make a mine out of a nickel property near Hope, B.C.

From left to right: Mr. Swasie (from Los Angeles), David Sloan, Victor Spencer, and promoter Ben Smith in front of a Junkers floatplane, Jericho Beach, Vancouver, May 1932. Photo courtesy of D. Sloan.

Mining aside, Victor Spencer's Pioneer stock was the salvation of David Spencer Limited, a family-owned chain of department stores in British Columbia. When the firm was struggling to survive in the depths of the Great Depression, the Bank of Montreal, who handled their account, decided it had had enough and unexpectedly cut off all credit and refused to honour the firm's cheques. Following a hurried meeting of the founder's five sons, Victor Spencer stormed into the office of the bank's western manager, Andrew Lang, the hard-nosed, irascible Scot who had harassed Dave Sloan a few years earlier, dumped the returned cheques on his desk and told him in no uncertain terms what he thought of the bank and its employees. Spencer caught a train to Toronto the same night and, by pledging his Pioneer stock, was able to get a rival bank to take the store account, a harmonious association that would continue until the Spencers sold out in 1948.

## *Bralorne*

Bralorne stock never appears to have had a daddy like Pioneer's Ben Smith and, as a poke at Pioneer, Bralorne in mid-1934 was proclaiming itself "a truly Canadian enterprise" with 91 percent of the stock held in Canada and 78 percent of the latter in British Columbia.[16]

If there was a daddy behind the scenes it may well have been Austin Taylor himself playing a lone hand. Bralorne was his mine. More than 300,000 of the one million shares outstanding had been issued to him, mainly through Bralco but also in exchange for Lorne Gold shares he had acquired along the way. Taylor was free to do as he liked; with Bralorne stock not traded in the United States there was no government body comparable to that country's SEC to keep track of and report on insider trading.

In general, Bralorne Mines gave out little information aside from printed annual reports and verbal statements at annual meetings. However, trading patterns suggest that with good news ahead, share prices were run up through timely press reports based on informal interviews with directors or unnamed sources. In contrast, bad news coming out of the blue could put Bralorne stock into a free fall.

Bralorne stock began trading on the Vancouver Stock Exchange in early September 1932 following the decision of the Lorne Gold Mines shareholders on August 5, 1932, to distribute the 400,000 Bralorne shares held by that company on the basis of one Bralorne share for every eight Lorne Gold shares held. On September 10, 1931, the first day with any volume,

650 Bralorne shares traded between 79¢ and 79½¢ a share. In late November the stock, until then traded in small volumes between 81¢ and 50¢, picked up and, in a remarkable week ending December 28, 1932, closed at $1.45 with a more than ten-fold increase in volume to 56,583 shares. The cause of the surge, discovery of spectacular ore in the King Vein west of the No. 1 Fault, must have been rumoured although, as yet, there was no official comment.

Bralorne stock, now listed in Toronto and Vancouver, rose steadily. In late May 1933, just prior to the release of the annual report with news of the new find, the stock topped the $8 mark on a combined seven-day volume of about 60,000 shares. Then in late July it hit $10 on a seven-day volume of more than 130,000 shares of the million share company, with over 114,000 of the trades on the Toronto Exchange. In mid-September, after a year of trading, it reached its 1933 high of $13.25 on the Toronto exchange, only to drop back and close the year at the $10 level.

Bralorne's shareholders approval of the Bradian Mines Limited spinoff at the January 4, 1934, meeting does not appear to have affected the price of Bralorne stock. The shareholders present probably paid more attention to Austin Taylor's statement that "within the next six weeks the directors would make a statement which should be pleasing to all stockholders."[17]

Bralorne stock, back at the $13 level in March 1934 after declaration of an initial dividend of 12½¢ a share, rose to a high of $17 in mid-July following news that the dividend rate had been fixed at 15¢ quarterly and the suggestion that more still might be paid in the form of bonuses. During much of June and July the combined volume of trading on the Toronto and Vancouver exchanges totalled 15,000 or more shares a week, more than twice the usual volume, with close to 80 percent of the action in Toronto. Trading in Bradian shares, now in the $3 range, showed a different pattern with the total volume only slightly less but more than 60 percent of the trades on the Vancouver exchange. Bralorne shares, following the $17 high, dropped slowly on lower volumes to close the year back at the $13 level, despite an extra dividend of 20¢ a share paid in late December to bring the 1934 total to 62½¢ a share.

Payment of the extra dividend made no sense unless it was intended to support the price of Bralorne stock. Recent developments at the mine had been disappointing, and it must have been obvious to the directors that the money could be better used either for an expanded exploration program or to ensure continuity of the regular dividend.

The first warning that all was not well at Bralorne came in a statement by Ira Joralemon dated March 26, 1935, and included in Bralorne's 1934 annual report:

> ... In the past few months there has been a drop in the grade of ore found and mined. This was to have been expected, as no mine could hope for continued spectacular developments like those in the preceding year and a half. There is no reason to be discouraged by the leaner period.... Reserves of developed and probable ore at the end of 1934 are estimated at 300,000 tons. The grade is uncertain. In all Bralorne orebodies, a considerable proportion of the gold is in plums of very rich material irregularly scattered through the leaner quartz. When the stoping happens to find many rich plums, the grade increases to 0.6 or 0.7 ounces per ton. Up to the middle of 1934 so many of these rich areas were found that it seemed probable that a 0.6 ounce average grade could be maintained. This now seems too high. When high grade spots happen to be widely separated, as in the past few months, the grade drops to 0.35 ounces.... All that can safely be said about the grade of ore reserves is that it is certainly as much as 0.35 ounces per ton, and the scattered rich areas will probably make the mill heads average considerably more than this.

About a hundred shareholders attended Bralorne's fourth annual meeting in the Hotel Vancouver on April 29, 1935. *The Vancouver Sun* reported:

> After the directorate and auditors of Bralorne had been unanimously re-elected, President Austin Taylor announced that officials of the company were prepared to answer any questions that might be asked and, in view of rumors which had been prevalent of late, he hoped the questions would fully cover the entire situation.
>
> In answer to a shareholder who intimated that recent market action was indicative of inside selling, in spite of the fact that the annual report showed 300,000 tons of ore in sight, Mr. Taylor said that no official of the company had ever advised anyone either to buy or sell shares, but confined their duties to a truthful, unvarnished statement of facts. They were not interested in market action.[18]

An interesting statement but, if the *Sun* reporter got Taylor's words down correctly, not much of an answer to the real question. Taylor and other insiders may not have been advising anyone but just what had they been up to themselves?

On the day of the annual meeting, Bralorne stock, which had been dropping steadily since the beginning of the year, closed at $7.35 on the Vancouver Exchange. There was more to come. Bralorne shareholders who had overlooked the first warnings were shocked to attention on June 6, 1935, by news that the company would not pay the 15¢-a-share quarterly dividend expected on July 15. More than 45,000 shares traded in a seven-day period, with the price falling from the $6 range to a low of $4.45. Three reasons were given for the directors' decision: the lower grade of ore being mined from the deeper levels of the King Mine, production lost during a two-week strike in May 1934, and the need for funds to carry out an extensive exploration program.

Writing in next day's *The Vancouver Sun,* mining editor Sidney Norman blamed the decline on heavy selling in Toronto, beginning early in 1935 after a eastern brokerage house had retained an engineer to examine the property for them. His conclusions had never been made public but from what Norman had been able to pick up:

> ... The report, while admitting the great possibilities of the mine, expressed the opinion that heavy expenditures would be advisable to place the workings in shape for maximum economical handling of over 400 tons daily to which the mill has been increased.

The engineer referred to was almost certainly John "Turn 'em Down" Reid, a consulting engineer in his late fifties and, by reputation, so close-mouthed that it was impossible to pry information about a private examination from him. Reid's Bralorne assignment in late 1934 or early 1935 may well have been prompted by the extra dividend of 20¢ a share paid on December 20, 1934. Courtney Cleveland, the young geologist charged with showing Reid around, recalled a somewhat hunched, tough-looking, little man with penetrating eyes who had to see everything for himself. Above ground, Reid spent days in the office going over assay maps and making his own calculations of the tonnage and grade of ore reserves. He had caught Bralorne at a bad moment, with the King Mine in trouble and, as yet, precious little to show for the Bradian work. Reid, in the front ranks of his profession, came by his nickname honestly, with most of the properties he rejected deserving their fate.[19]

The manager of the Bank of Montreal's Bralorne branch, taken on a four-hour underground tour on an evening in mid-December 1934 and assured by the staff member guiding him that Bralorne had ten years of ore in sight, had second thoughts in a diary entry for July 22, 1935:

> I hope they are not going to gut [Bradian] as they have apparently done to Bralorne. Am certainly losing faith in the Bralorne directors, especially the president, whom, I understand, sold about 155,000 shares last fall when the much disputed dividend bonus was paid. I am afraid that Austin is too smart for the boys.[20]

Bralorne's bonanza, soon to be known as the King Mine, had indeed been gutted. But the company had a future in its subsidiary, Bradian Mines Limited, spun off in January 1934 and hastily recombined in late July 1935. In late September 1935, Bralorne shareholders were mailed an extract of a report by Joralemon, now a Bralorne director, recommending that a reserve of half a million tons of indicated ore and a million dollars in the treasury be built up before resuming dividends. This would provide a safety factor for a lean period such as the company had just gone through and, from all indications, the goal would be reached within six months. It was, and dividends resumed on April 15, 1936, with a payment of 15¢ a share, described as 10¢ on a regular quarterly basis plus a 5¢ bonus. Bralorne prospered and for nine more years dividend payments were far in excess of the proposed rate. Bralorne stock took longer to recover and traded in much smaller volumes. It would drift below the $10 mark until some time in 1938.

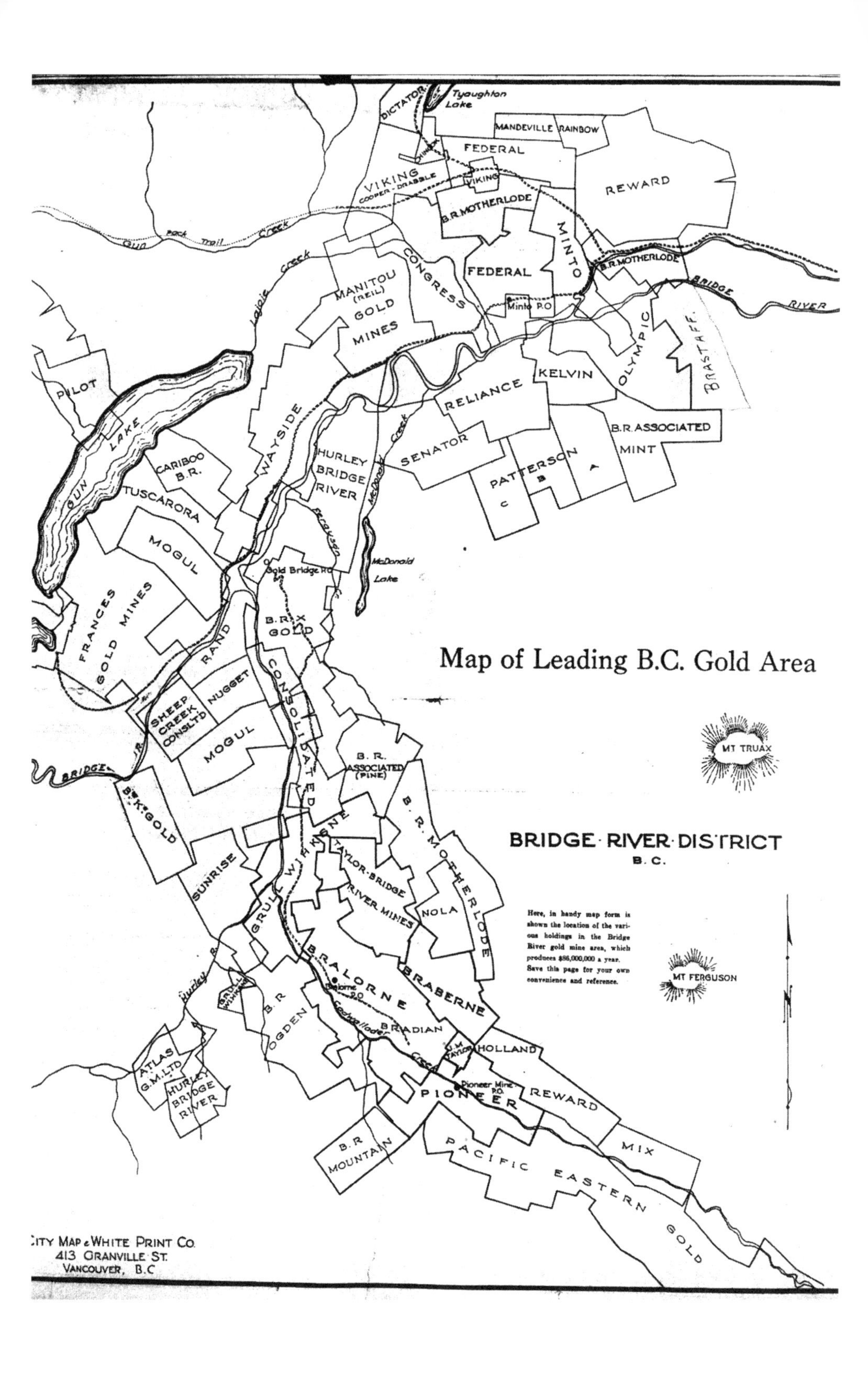
Map of Leading B.C. Gold Area
BRIDGE·RIVER·DISTRICT
B.C.
Here, in handy map form is shown the location of the various holdings in the Bridge River gold mine area, which produces $86,000,000 a year. Save this page for your own convenience and reference.
Tyaughton Lake
DICTATOR
MANDEVILLE
RAINBOW
FEDERAL
VIKING
COOPER - DRABBLE
REWARD
B.R.MOTHERLODE
MINTO
FEDERAL
CONGRESS
MANITOU (REIL) GOLD MINES
Minto P.O
B.R.MOTHERLODE
BRIDGE RIVER
OLYMPIC
BRASTAFF
KELVIN
RELIANCE
B.R.ASSOCIATED
MINT
SENATOR
PATTERSON
A
B
C
Gun Creek
Pack Trail
Lajoie Creek
PILOT
GUN LAKE
CARIBOO B.R.
TUSCARORA
WAYSIDE
HURLEY BRIDGE RIVER
McDonald Creek
Ferguson Cr.
MOGUL
McDonald Lake
Gold Bridge P.O
FRANCES GOLD MINES
B.R.X GOLD
RAND
CONSOLIDATED
NUGGET
SHEEP CREEK CONSL'T'D
MOGUL
BRIDGE R.
B.R. ASSOCIATED (PINE)
B.K.GOLD
WIHKSNE
B.R. MOTHERLODE
SUNRISE
TAYLOR-BRIDGE RIVER MINES
NOLA
GRULL
BRALORNE
Bralorne P.O.
BRABERNE
Hurley R.
GRULL WIHKSNE
B.R. OGDEN
Cadwallader Creek
BRADIAN
J.M. TAYLOR
HOLLAND
ATLAS G.M.LTD
HURLEY BRIDGE RIVER
Pioneer Mine P.O.
PIONEER
REWARD
B.R. MOUNTAIN
MIX
PACIFIC EASTERN GOLD
MT TRUAX
MT FERGUSON
CITY MAP & WHITE PRINT CO.
413 GRANVILLE ST.
VANCOUVER, B.C

# *High-Grade and High-Grading*

To miners the word "high-grade*" has a double meaning: as a noun it refers to very rich ore and as a verb to the theft of such ore and its illegal sale.

Through the years, the grade of ore milled at Pioneer and Bralorne averaged just over half an ounce of gold per ton plus a lesser amount of silver. The precious metal alloy, technically known as electrum,* is much heavier then the enclosing quartz and, for half-ounce ore, a comparison by volume works out to roughly one part of electrum in 250,000 parts of quartz. In much of the ore, gold was present as tiny grains invisible to the naked eye, but here and there pockets of high-grade studded with coarse gold were found, especially at Pioneer.

In management's possession, high-grade might be either amalgamated with mercury or stored in a vault as a handy reserve that could be used to boost a low monthly production figure. Found by a miner or a mucker, it might simply be glanced at and tossed in with the rest of the ore, but there was always the temptation to smuggle it out of the mine, either to keep as a specimen or to sell. In Canada, high-grading was considered theft, unlike the bonanza days in the mines of the American west where, unofficially at least, it was shrugged off as one of the costs of doing business.

In the spring of 1937, spectacular high-grade was struck at Pioneer. There was nothing to suggest its presence; the discovery stope had been started above 10 Level in the Footwall Vein following a wide section of the vein carrying low gold values. They were, so low, in fact, that Dr. James, now managing director, had ordered Bob Eklof, his mine foreman, to stop working it as there was better ore elsewhere in the mine. On protesting, Eklof was given a choice, either follow orders or prepare to "roll his blankets" and move on. Eklof, just as determined, ordered a section of the stope drilled and blasted one Friday when he was short of broken ore needed to keep the mill operating over the weekend. After the blast, the working face was hidden beneath an unusually heavy coating of black dust composed mainly of pulverized sulphide minerals. Washing the dust off with a water

*Opposite:* A map of claims in the Bridge River area, from a special report in the *Vancouver Sun* of August 19, 1936, illustrating how heavily staked the valley was.

hose revealed a vein, six to eight inches wide, laced with solid gold.[1]

When James returned to Pioneer the following Tuesday, Eklof wasn't dressed for underground and when questioned, replied that he had continued to work the stope and was getting ready to roll his blankets. James, unaware of the find, was persuaded to go underground for a quick look. Eklof stayed and soon after, the drift leading to the stope was closed off with a locked steel gate, and two specially bonded miners worked there for a year. The high-grade, continuous to 9 Level about 120 feet above, was mined carefully, using light charges to blow it down onto canvas sheets where the richest material was hand sorted and packed in powder boxes for the trip to the surface.

Also at Pioneer, there was joking about streets of gold, especially after a small but spectacular find was made by a hoistman on the road in front of his house. He appeared at the mine office on Sunday morning and handed Dr. James a dish-shaped piece of quartz about the size of a small pie plate with free gold running through it, kidding that it was a lovely specimen but a bit too large or obvious to keep on the mantelpiece. Apparently it had been lying on the road for several years and must have been turned over by a passing car or truck, exposing the gold. Waste from an exploratory crosscut, consisting mainly of barren wallrock, had been used on the road and the high-grade was believed to have come from a very narrow, high-grade vein cut while driving it.[2]

Bralorne too had its high-grade. A geologist working in the King Mine in the mid-1930s recalls examining a freshly-blasted working face and finding a portion of it rich in silvery arsenopyrite laced with tiny veinlets and blebs of gold. He and a sampler working with him collected about 150 pounds of the best material and carried it to the mill where, on treatment, it was estimated to have yielded about $10,000 worth of gold, close to $70 a pound! A good day's work for the pair who, between them, probably earned less than $20 a day.[3]

## High-Grading

Everyone working in the mines was conscious of high-grading and, though probably not involved, might well have doubts about some fellow workers. Certainly there were markets for the stolen gold, but with it went the chance of getting caught and ending up in jail. Two groups were supposed to be on the lookout for high-graders, the Provincial Police and the salaried staff at the mines. But one young constable had his doubts about the staff's concern when on a visit to a mine manager an unlocked desk drawer was pulled open to show off four or five spectacular specimens kept there.

In general, mine staffers kept an eye on things but left actual investigations up to the police. Miners coming off shift were not subjected to a body search or even required to open their lunch boxes for inspection. Sometimes a subtle warning was all it took to discourage a would-be high-grader. One engineer, on moving a miner's lunch box out of the way, noticed that it was suspiciously heavy and, on opening it, found it full of high-grade. Rather than confront the owner he replaced the ore with a handful of spikes of about the same weight and threw the goodies down a nearby ore pass.*

The sections of the mills where gold and amalgam were handled were partitioned off and kept locked, with access restricted. After the gold had been cast into bricks of about 60 pounds, weighed and stamped with identifying marks, it was taken to the post office where it was signed for and shipped by registered mail to the mint in Ottawa. No armed guard rode

Pioneer gold bricks, each weighing about 60 pounds, and high grade gold ore, 1934. Leonard Frank photo 17300, courtesy of B.C. and Yukon Chamber of Mines.

shotgun on the mail run to Shalalth but the risk was probably minimal as, unlike crude gold, a stolen brick with its distinctive fingerprint of trace elements would be difficult to dispose of.

One of the few scares occurred on a Saturday afternoon when a mail sack containing a gold brick properly packaged and with postage attached disappeared from the Bralorne post office. The sack, ready to ship on Monday, was last seen on a small counter near the door of the post office, and it was thought that the relief driver on the Evans freight and mail truck might have picked it up by mistake. The police were called in and, with the news kept quiet, postmaster C.P. Ashmore and the police had an anxious wait until the driver, off playing in a ball game, came back to Shalalth, returned their phone call, and confirmed the safe arrival of the brick.[4]

There are many legends about high-grading, some still repeated today, often by those claiming to have "inside" information. According to one, the mining companies employed men in Vancouver to buy high-grade in local beer parlours, finding it as cheap to recover their gold this way as to mine and mill it themselves. Another story had a Bralorne "nipper*" whose chores on the last shift of the week included emptying the "honey boxes," wooden powder boxes used as latrines, concealing high-grade in his fragrant cargo and caching it somewhere on his way to the dump. Yet another was that much high-grade was smuggled out of the Bralorne mine through a ventilation shaft that reached the surface close to Mrs. Noel's cabin on the old Lorne workings. Still others refer to stolen gold being passed off as production from certain Cariboo placer mines, its origin concealed by amalgamation with mercury and retorting to yield sponge gold, a common product of placer mining. Then there was speculation about the local Chinese: perhaps they were the front men for a Chinese network with its own, secret uses for unlimited amounts of illegal gold. It was a new twist to an old idea. In his 1895 report, Lillooet's gold commissioner had noted, "From the white [placer] miner and the Indian a fairly truthful account of their doings can always be had, but from the Chinese it is very different and . . . there is a large amount [of gold] unaccounted for . . . an amount that will never be known."[5]

The legends are more colourful than the few cases where high-graders were arrested and brought to trial, with one notable exception in the post-war period described later in the book. But no one really knows the losses, and possibly the most certain thing is that any high-grader who got away with it must have been very close-mouthed at the time and is unlikely to be saying much, even today. As "Ma" Murray put it, the odds were against the married high-grader, "because his wife gets in the habit of telling. But a bachelor highgrader is said to be immune from detection ... because he can pound out gold at night without anyone looking over his shoulder or nagging him to go to bed or to play cards or go to the movies."[6]

## Pioneer, November 1932

A high-grading case involving three former Pioneer employees was tried in Vancouver's police court after a raid on one of the men's home discovered a complete mini-plant for the recovery of gold. The trio pleaded guilty to retaining gold stolen from the Pioneer mine, but their counsel, pleading for leniency, claimed they had yielded to temptation and recovered gold from furnace linings, crucible bricks, sweepings and other materials stored in barrels beside a road at the mine and considered waste. Dave Sloan, called as a witness, testified that the material was not looked on as waste and that the gold recovered, some $6,800 worth, could not all have come from this source. The magistrate sentenced the mine's former assayer, whom he considered the ringleader, to two years and five months in jail and his accomplices to shorter terms.[7]

## Pioneer, April 1936

The belief that Pioneer employed a man to buy back gold stolen from their mine may well arise from a police raid on the Gold Bridge drugstore when gold worth $1,500 was seized and the druggist and two others charged with possession of stolen gold.

At the jury trial held in Kamloops that June, it came out that the trio had been trapped by a former Vancouver detective, Alex Imlah, then working for Pioneer. In October 1935, Imlah had approached the druggist, whom he had known for several years, seeking to buy gold. Put in touch with a second man, a cook at the Wayside property, nothing had happened until the following April when Imlah had been offered the chance to buy "retort" gold. After some haggling, the former detective offered to pay $1,000 for ten pounds of the material and made a telephone call to a "Mr. Balderson" in Vancouver to have the funds wired to the Gold Bridge bank. It was sloppy work; both the druggist and the cook recognized the number as that of Victor Spencer, president of Pioneer Mines. Despite that, they accepted the thousand dollars on Imlah's return about an hour later.

The gold had come from an employee in the Pioneer refinery, paid a wage of $150 a month to handle $225,000 of the mine's gold each month. He now pleaded guilty and, after testifying for the Crown, received a nine-month suspended sentence. A jury found the other two not guilty, apparently agreeing with the defence counsel's suggestion that Imlah, acting for Pioneer and using their money, could be considered an agent provocateur.[8] If nothing else, the case was a warning to other would-be high-graders that Pioneer, concerned over gold losses, was on the lookout for them.

## *Bralorne, February 25, 1939*

George Gordi, a Bralorne miner spending Saturday night at a Lillooet hotel, answered a knock at his door to be confronted by two policemen who, tipped off by a source they refused to divulge, had been keeping an eye on him. He had certainly been acting suspiciously. He appeared to have a large sum of money and, after having his hair cut and doing a few other things, he had retired to his room, apparently planning to take the gas-car to Shalalth early the following morning. It made no sense, why make the trip to Lillooet if that was all he had in mind? Surely he must be up to something. The local magistrate, agreeing with the policemen's suspicions, had issued a search warrant.

Gordi had close to $500 in cash and, after questioning him for an hour and a half, the police talked him into turning it over to them and spending the rest of the night in the lockup although, as yet, he had not been charged with anything. Next, the police went on to question Jim San, owner of Jim Brothers general store in Lillooet. It took until 6:00 next morning to get the Chinese to open the store's safe. On searching it, there was no sign of illegal gold. However, by early afternoon Jim San must have had second thoughts about it all, turning over to the police a glass sealer jar containing 22½ ounces of crude gold worth an estimated $660.50 and admitting to paying Gordi $505 for it.

On Monday the police escorted Gordi back to the Bridge River and searched his cabin at Ogden. They found pieces of high-grade and, subsequently, traces of gold in the flour sifter, on hammer faces and even in Gordi's clothes. The source was a puzzle: the accused, one of the most respected miners in the camp, worked mainly on a contract basis driving raises through waste rock. "Quiet and reserved, he took his part in the social life of the camp and no one was more surprised at his arrest and ultimate charges than his nearest neighbors."[9]

Charged with theft of and trafficking in gold, Gordi was defended at the preliminary hearing and subsequent trial by Angelo Branca, a rising criminal lawyer from Vancouver who eventually became a judge in the Supreme and later Appeal Courts of British Columbia. Bralorne, in turn, retained Jack Nicholson, "a hundred dollar a day lawyer from Vancouver," to represent them. At the jury trial in Williams Lake, Branca concentrated on Jim San, the chief Crown witness who, once on the stand, seemed to develop a sudden inability to understand English and, on close questioning, a tendency to claim that he was feeling sick at the time and therefore unable to recall exact details. Branca persisted, refusing to call in an interpreter. The *Bridge River-Lillooet News,* not known for its sympathetic portrayal of Chinese, had already carried Ma Murray's report of an exchange at the earlier preliminary hearing:

> Angelo Branca, brilliant young Vancouver attorney, dark, flashing eyes, many impulsive gestures, goes into action. He roars at Jim.
>
> "Where did this man come from?"
>
> Jim: Blidge Rivah.
>
> Branca: Bridge River is a large district; where in Bridge River.
>
> Jim: Maybe Blalorne; maybe Pioneer; I don't know.
>
> Branca: You don't know? Have you ever seen gold that came from Bridge River, from Bralorne or Pioneer?
>
> Jim: No. I don't know where he come from so long as he is gold.
>
> Branca: How did you know this stuff was gold (pointing at the exhibit).
>
> Jim: Gold is heavy.
>
> Branca: This might be lead not gold.
>
> Jim: No. No lead not so heavy as gold.
>
> Branca: Then you have had no previous experience with gold from Bridge River valley.
>
> Jim: No. So long as he is gold. I buy him. I don't ask where people come from.
>
> Branca: If I came to your store then and offered gold, you could buy it and ask no questions?

Jim: Yes. I buy him and sell him.

Branca: How much did you buy say in January.

Jim: Maybe $50 maybe $100. Maybe very much; maybe verry little.

...

Branca: From whom did you buy in February?

Jim: Me buy Peter Di'ablo (an Indian named Peter the Devil, who can always produce a poke with a hundred dollars worth of gold which he wins from secret diggings on the Fraser River). Me buy Peter Di'ablo, Fred Thomson, George Gordi.

Branca: Do you know the law about reporting purchases of gold?

Jim: Me buy alltime before February and keep no record and make no report. Me make report now alltime

...

Branca: Do you swear you never before saw gold from Bralorne or Pioneer before?

Jim: Yes. I never saw gold from Bralorne or Pioneer before.

Pointing to prisoner, Branca said: Did you know George Gordi's name before you saw him at Ashcroft in the line-up?

Jim: Yes I see him in Blidge Livah-Lillooet News. I don't know his name when he come my place

...

Branca: What was there about the accused that made you remember?

Jim: When I see a man's face he can't miss him.

Branca: What about his face?

Jim: Same facee he have got now. He face yellow and a long, short facee. I know him he is the man that all he know. He have coat, black overcoat. Other men in line-up at Ashcroft, some short some long.[10]

Throughout the trial, Branca stressed that it was Gordi's word against that of the Chinese. The tactic paid off, for in Ma Murray's words:

> It was evident that the judge and the crown prosecutors had very little conception of the ways of the miner and the historical precedent of the Cariboo where Oriental evidence is concerned, because the trio were undeniably and shockingly surprised when [the jury] foreman stood at attention and said "We find the accused Not Guilty, Your Honor."[11]

That evening, after Branca conferred with the prosecution, Gordi reappeared in court and pleaded guilty to lesser charges arising from the gold found at his cabin. Granted leniency on grounds of his claim to be 67 percent silicotic, he was sentenced to six months in jail, half of it already served since the date of his arrest. The judge, disgusted with the jury, discharged them from further duty for five years and threatened to take away their voting privileges.

## *Pioneer, October 1940*

Discovery of a cache of high-grade may well have been a sequel to the 1936 raid on the Gold Bridge drugstore and Ma Murray pulled out all the stops in describing it for her readers in the *Bridge River-Lillooet News* of October 11:

> GOLD GOLD GOLD
>
> It was nothing but! Talking about Gold! Working with Gold! Morning, noon and night. Grubbing for Gold! Mucking for Gold! Mining Gold! Milling Gold! Pouring Gold! Stealing Gold!
>
> Lee Dunn the curly fair-haired young chef at the Pioneer Club Cafe, serves mostly miners who board with him and work for the big Pioneer. Gold is as foreign to Lee as pancake batter should be to a mucker. The September day was hot and the meals were numerous. The salt was lumpy and the milk was sour. In short everything had gone wrong. Lee decided he would run away out into the woods, all by himself - to meditate, maybe on the wrongs of life! . . . It was purely hookey or air cas-

tles. It was also getting away from gold-talk and gold-fever. But Lee couldn't get away from the filthy stuff.

Listen to this and let Ripley challenge it if he will. Sitting at the butt of a big Jackpine along the Cadwallader... Lee Dunn ground his heels into the gravelly earth bed. It was wet and moist. It crunched into a cup like hole. Shining in the daylight was a small nugget of the yellow metal. Lee gingerly reached down his hand between his two heels to pick up the pellet. He found a few more nuggets clustered around. He thought it was a necklace. He carefully lifted them out. More and more nuggets tumbled in from the sides of the small gravel-saucer. He scraped the gravel farther back and uncovered a literal pile of gold ore, pure nuggets and high grade rock. Imagine the surprise this young lad, in his twenties, had when the farther he dug into the ground the more gold he unearthed. How little regard he had for the metal was shown in the fact that he gathered up the nuggets and as much high grade rock as he could carry and went straight away to the Pioneer office where he threw down the cache on the desk of Dr. James. Imagine the surprise Jack Pearson and Phil Stevenson got. Imagine Ed Emmons, the supt. observe a pile of gold nuggets from practically out of the sky. Then picture if you can, Geologist "Bud" Rose bug his optics. Try and find sympathy for chef Lee Dunn who had run away from gold to run into it, to be delivering a large cache of evident high-grade, trying to explain where he got it, how he happened to be out looking in that spot and at that time.... Believe it or not Lee Dunn had uncovered a cache that was parked within a half mile of Pioneer main office in 1936 when detectives got hot on the trail during the purge of highgraders that concluded in a court action in Kamloops.... This cache is alleged to have been buried by the late Allan Cameron who was suspected along with a great many more miners when the Pioneer was rumored to be losing $10,000 weekly from thefts and an organized ring of highgraders who were at work. Allan Cameron was killed shortly after on the Cariboo Highway in company with another valley miner known as Oliver. Rumor has it that many old-timers knew that Allan had cached some high grade when it got too hot and had intended to come back for it....

The Pioneer company went up to the spot led by Lee and with shovels and powder boxes carried it away. Mum was the word when we enquired the amount of the find. Some said it was $3000 and others said it was $30,000. There were two powder boxes, a canvas bag and considerable in nuggets. Be that as it may or the ownership of the cache, anyway Lee Dunn has decided to settle down and be contented with the gold that is talked about daily by his charges. We do not know if the Pioneer awarded him for his honesty or if they suspected him. We did learn however that Lee has become owner of a fine new fishing rod and a benefit amongst the boarders is now being organized to present him a large basket. That the hopes that his cache of finny skill will be in comparison with his gold mining luck is evident. M.L.M.

*Overleaf:* Scab list, crudely mimeographed and with derogatory comments, posted by defeated Pioneer strikers, March 10, 1940.

SCAB LIST - PIONEER MINE STRIKE - PLEASE POST PROMINENTLY

March 10 1940

MINERS

Alex Stelmock - Hockey player from Merritt BC.
Alex (Goof) Neilson - Baseball player from Merritt BC, real rat.
James Neilson - Hockey player, Merritt BC, alleged scab herder.
Bob Saunders - Hockey player, moustache, sneaky features, 100% rat.
Don Bravender - Medium size and complexion, glasses, chief stooge.
Chas. Lessard - Former barber, short and dark complexion.
Wm. Nasadyk - From S.W. Alberta, fairly tall and thickset.
Wm. Wycherly - English accent, tall, thin and hungry looking.
Frank Prewet - English accent, short, heavyset.
Jack West - Fairly tall and heavy, (scabbed under pressure)

TIMBERMEN

Wm. Cassidy - Hockey player from Merritt BC, red face.
Joseph Muller - "Nazi Joe", German Army man, tall, heavy set.
Ed. Diffner - Can. Swede, fair, from Matsqui BC, top stooge.
Chas. Plant - Middle aged, stout, formerly Sandon BC.
Magnus Tate - Hockey player from Merritt BC, tall, fair.

MUCKERS

Karl Donner - German fascist, plays piano, medium complexion, tall.
Fred Ritson - Carpenter, short, middle aged, medium complexion.
Wm. Pearce - English, rather short, first class stooge.
Wm. Wysocki - Pole, short and dark, very stupid.
Syd Bridson - English, tall and thin, good stooge.
Frank Millard - Tall and simple, medium complexion.
R. Falconer (Scotty Marks) - Short wrestler, choir singer, 100% rat.
Bert Hazel - English, medium size, fair, rabbit type.
James Schrump - Hockey player from Rossland BC, short and timid.
Geo. Coghill - From Tantallon Sask., tall, dark, well built.
Mike Tarentiuk - Walking skeleton, speaks poor English, middle aged.
Mike Crowe - From Coal Branch Alta., hook nose, dark, stool pigeon.

HOISTMEN

Adam Broomfield - Medium height, sandy complex., double crosser.
Wm. Matthews - Tall, prematurely grey, dark complexion.
Shorty Olsen - Dane, short, stout, middle aged.

STEELSHOP

John Patsula - Tall, fair, natural stool pigeon.
Reg. Slater - Medium height, fair complexion.
Harry Gough - Short, lean, rat type.

MILLMEN

Dick Winch - Habitual scab, young, tall, dark, 100% rat.
John Anderson - Formerly Anyox, tall, grey haired, well built.
Gorden Davies - Medium size, dark, baldish, former railroader.
Art Prosser - Medium size, receding hair, dark.
Wm. Leverett - Short, undersized, real rat type.
Beaton Patience - Brown hair and eyes, too young to know better.
Noel Peters - Tall, slim and slippery, reporter for Vancouver Sun.
Ed. Motte - Belgian, medium size, fair, blue eyes.
Pete Broomfield - Young, short, fair, (scabbed under pressure)

SURFACE

James Lamb - Machine doctor, tall, glasses, real stooge.
Bob Gracey - Machinest, middle aged, glasses, short, thickset.
John Ruski - Young, wavy hair, dumb.
Jas. King - Old man, union minded but mislead.
Jas. Duncan - Grey hair, medium size, carpenter.
Alex Broomfield - Young, fair hair, short, mislead.
Wilf. Maurice - French Can., dark, medium size, rabbit type.
Ken. Milne - Young, wavy hair, medium complexion, dumb.
Joe Bertoia - Italian, short, thickset, dumb.
Vern Clippingdale - Tall, slim, glasses, choir singer.
Jas. Milne - Tall, well built, painter, medium complexion.
Archie Davies - Tall, slim, receding hair, electrician.
A. Buhler - Tall, dark, loose jointed, from States, real stooge.
Geo. Jones - (Red) Gardener, choir singer, red hair.

SHIFTERS

Lyle Knight,
Ralph Futcher,
Ivan Novasol,
Mark Hamlund,
Tom Bevister, (mill)
Mike Wysocki,
John Dragvik, (mill)

# *Strikes Won and Lost*

By 1935, the miners were restless, impatient for some of Pioneer and Bralorne's good fortune to rub off on them. Resentment still smoldered over a 25¢-a-day wage cut imposed by Dave Sloan a few years earlier and never rescinded. Yet both companies were paying dividends and had received a windfall in the price of gold, boosted from $20.67 to $35 US in less than a year. Caught up in the excitement of the Bridge River boom, it was all too easy for the miners to forget that jobs were still scarce elsewhere in the province and that base metal mines were struggling to stay open in the face of depression-level prices for their products.

There was no union to organize a protest. At Pioneer a delegation of the workers approached Dr. James asking for a 20 percent wage increase and on being turned down by management the men went on strike Saturday, May 4, 1935. At Bralorne, the men assembled in the community hall and, after an orderly discussion, voted to strike if their demand for a 20 percent wage increase was turned down. The vote was 105 to 98 in favour, most of those opposed being married men with families in camp who worried that a long or lost strike could mean a difficult relocation.[1] Despite the close vote, there was solid support for the strike after the company refused their demand, and the men went out on Tuesday, May 7, 1935.

After some initial huffing and puffing and an aborted attempt at mediation by the province's Deputy Minister of Labour, the companies realized that, union or not, they had a real strike on their hands that would last until they came up with a reasonable offer. It took just over two weeks to settle. When the strike ended on May 21 miners' wages were $5.40 a day, up from $4.75; muckers' $4.50, up from $4.00; and surface labourers' $4.00, up from $3.25.[2] In addition there were minor concessions such as reduced prices for work clothes in the company stores. There had been no trouble and the day the strike ended a group of 23 provincial police, held in reserve at Craig Lodge on Seton Lake, started back to Vancouver. Some six weeks later Dave Sloan, as quoted in *The Evening Province* of July 6, 1935, seemed to be shifting the blame to the government's liquor policy:

> I don't want to keep any man from his drinks, but when miners alternate between mine and beer-parlor, dissatisfaction is bound to follow, leaving the way clear for the work of Com-

> munist agitators. Government exploitation in the Bridge River country is the direct source of the recent strike trouble, not poor labor conditions or an unfair wage scale as claimed.

About this time the mines switched from a seven to a six-day work week. Where feasible, Sunday operations were shut down, and in the mill, a continuous operation, shifts were adjusted so that each worker had four days off over a 28-day period. There was still no attempt to organize the mines.

Early in 1936 employees and management at Bralorne formed a co-operative committee similar to one at Consolidated Mining and Smelting Company's smelter at Trail, B.C. The committee was composed of an average of 16 members, elected by employees for one-year terms to represent the departments of the mine; the number of members from each depended on the number of its employees. Management was represented by the general manager, a non-voting member. Grievances and other matters of concern were discussed at semi-monthly meetings, and a safety sub-committee made regular inspections of the mine, reporting back to the full committee.[3] For the moment at least, both sides were satisfied with the arrangement. Pioneer employees hung back, but in late October 1938, voted to organize a similar co-operative.

There was more trouble at Bralorne in late March 1939 when the company announced a restraint program to begin on April first. It called for senior miners to take a compulsory holiday of 30 days during the coming months and, by doing so, protect some of the junior employees from permanent lay-off. The men were opposed and, after considerable discussion, voted 337 out of 413 to continue negotiating with the company through the co-operative, rather than through the Bralorne Miners Union, Local 271 of the International Union of Mine, Mill and Smelter Workers.[4] The Mine-Mill local, organized in December 1936 and itching to get involved, was prohibited by legislation from doing so until it could demonstrate support by a majority of Bralorne employees. As for the restraint program, it appears to have been forgotten.

## Pioneer Strike, 1939–40

At Pioneer, things were different, and there was less give-and-take between management and employees. Perhaps personalities were involved but, whatever the reason, problems that might well have been resolved were allowed to grow into an angry confrontation. By 1939 there was no turning back and both sides became pawns in an all-out struggle to organize the province's mines. As at Bralorne, the Mine-Mill union was involved. Affiliated with the CIO (Committee, later Congress, for Industrial Organization) the union was anathema to the mine operators, who considered it communist-led.

Trouble began in April 1939 when Howard James, managing director of Pioneer, turned down a request by the miners for paid vacations, stating that the company was using its surplus funds to search for new mines that would employ more men. Three organizers from Mine-Mill appeared in camp about this time and, at a meeting held later in the month with roughly 60 percent of the hourly-rated employees present, the men voted 135 to 24 to join the union.[5]

Fitful negotiations over recognition of new Local 309 followed. Under provincial labour law, as amended in 1938, an employer was not required to bargain with a union as such, unless a majority of employees were members prior to December 8, 1938. Unions signing up a majority after that date were obliged to call a meeting of all employees and elect a bargaining committee by secret ballot. The committee, although it could be the same as the union executive, then dealt with the employer on behalf or all employees, not just union members.

In early June 1939, a dispute arose over the dismissal of a union member, and an official of the province's Industrial Relations Board was sent to Pioneer to assist in resolving it. After checking the union's membership against the payroll, the official negotiated between management and the union executive, bringing about the employee's reinstatement.[6] In effect, the union had won recognition of a sort from Pioneer, but the company, stalling for time, had no intention of acknowledging it until forced to do so.

After letting things drift in the summer, the union renewed its campaign for recognition in early September. The executive, told by Dr. James that he would co-operate with them only as far as compelled by law, called a meeting for September 4. At it, union recognition, the check-off for union dues, and a general wage increase of a dollar a day were called for. A vote taken by secret ballot was 125 to 17 in favour of striking, if government conciliation failed to bring about a settlement. A telegram to the provincial Minister of Labour requesting appointment of a conciliator resulted instead in the arrival of M.H. McGeough, a departmental investigator lacking authority to deal directly with the disputants.[7]

McGeough called for a general meeting of employees to elect a bargaining committee, a proposal tentatively accepted by the union executive but when put to the meeting rejected on the grounds that the union committee had been recognized as such in the dispute in June. As an alternative, the union executive drew up a petition that the men could sign as either in favour of or opposed to the present committee.

Nothing was resolved and confrontation came on Sunday, October 8, 1939. That day a mass meeting repeated the three demands, and then, upset that ten days had passed without the appointment of a conciliator, turned it into an ultimatum with the strike called for 6:00 that evening if Pioneer management refused to negotiate. Dr. James, given less than an hour for his answer, refused to meet with them and the strike was on.[8]

McGeough then posted a formal notice calling for a meeting the following afternoon "to nominate and elect a committee to bargain collectively as representatives of all employees of the Pioneer Gold Mines of B.C. Ltd." But, at the meeting, he announced that his latest instructions from the Minister of Labour in Victoria were that the strike was illegal and, therefore, the men were no longer employees of Pioneer. If they wished to vote for a bargaining committee they could only do so after returning to work. Dr. James proposed that everyone, both union and non-union, return to work at 8:00 that evening. Supporters of the Pioneer co-operative agreed at once but the union members rejected the offer. By now, the union executive realized they had acted hastily, because both government legislation and the parent union's constitution called for a 14-day waiting period after all preliminary procedures had been complied with. Be that as it may, the executive had made its stand and it was too late to turn back.[9]

The provincial government made the next move. It charged six members of the union executive with (a) striking before applying for a conciliation commissioner and therefore illegally; and (b) refusing to bargain collectively with the company through representatives duly elected by majority vote of the employees. Both actions were contrary to the Industrial Conciliation and Arbitration Act, a relatively new piece of legislation. The six were summoned to appear in the Bralorne police court on October 24, 1939, but, at the request of their lawyer, John Stanton, the case was remanded until November 1. Years later Stanton wrote of the trial:

> The first trial, presided over by an elderly farmer named George J. Sumner, was held in the Ace Dance Hall at Gold Bridge. The structure was liberally decorated with pin-up girls and packed with interested spectators. After two days of hearings, William Cameron, the union's president, was convicted of striking illegally and was fined $150 or in default, three months in prison.

> Although it was not my first experience with a client being jailed for union activity, the decision seemed so monstrously unjust that I must have shown emotion as I gathered up my papers.
>
> Along came a young fellow who clapped me on the back, shook my hand and . . . "Cheer up,"
>
> "Whatever for?" I asked, wondering if he was putting me on and concluding he was not. "One good union man, totally innocent, faces jail because the union can't pay his fine. Five others are sure to follow unless there's a miracle. A strike can be lost. What's to be cheerful about?"
>
> "Think nothing of it," rejoined the young man. "That old bugger (pointing to the just-vacated podium) is my uncle. We all know he always finds everyone who comes to his court guilty."
>
> "How so?" I asked, intrigued by the idea.
>
> "It's simple. My uncle says he has to find 'em guilty because the police will never take anyone in unless they *are* guilty."
>
> It took me a moment to absorb this revelation about the administration of justice in the Bridge River Country.
>
> "But what can people here do when your uncle jails them when they know they're innocent?"
>
> "Oh, they just appeal to Judge Wilson in Lillooet and he lets them go. You should try it."
>
> "I certainly will." And that, later, is what I did.[10]

There was a contretemps next morning when the remaining five failed to appear in court. The defence claimed that by failing to adjourn their cases, an error in procedure, the court had lost the jurisdiction to try them. The prosecutor, caught off guard, demanded and got bench warrants for the missing five. All were taken into custody and had to be bailed out. In the meantime, Stanton succeeded in getting a Supreme Court order forbidding Sumner from proceeding with the cases, but by the time the official telegram reached Gold Bridge Sumner had already finished, fining them a total of $775. As predicted, subsequent appeals to Judge Wilson of the Cariboo County Court cut the fines by two-thirds. All were paid and no one went to jail.[11]

Not all Pioneer miners supported the union. The non-union men held their own meeting in mid-November and the hundred-odd attending voted 96 percent in favour of returning to work. A repair gang of 25 men was sent underground but a picket line went up and James, deciding not to risk violence, laid the gang off again at noon the next day.[12]

There was a potentially dangerous incident when two boys, both sons of senior mine officials, planted a homemade bomb beneath the window of a miner's house where John Stanton could be seen working at his papers. The bomb, a mixture of sulphur and zinc dust in a glass jar, went off prematurely and the blinding flash silhouetted the fleeing culprits against the skyline. Fortunately, Stanton was not hurt, although the bomb blew in the window, set the curtains on fire, and burned holes in the copy of the Industrial Conciliation and Arbitration Act he was studying.

For the moment it was a stalemate. The company was unable to reopen the mine, and the union members held out in the bunkhouses and some of the company houses. Early in December 1939, the company served eviction notices on 35 families at Pioneer, but the union got around this by swapping houses, frustrating the deputy sheriff's attempts to match up houses and occupants. Initially Bralorne miners had helped the union, raising $1,700 at a payday in October but, by now, they were beginning to waffle. At a mass meeting held early in December they voted 316 to 131 to stick with their co-operative and defer wage demands in favour of a bonus system.[13]

Evicting the financial secretary during the Pioneer strike of 1939-1940. Photo courtesy of Irene Howard.

During December 1939 there were negotiations behind the scenes, in part involving Harold Winch, an opposition member of the legislature in Victoria. As Winch understood it, Pioneer would reopen in the New Year employing all those on the payroll when the strike began, with no discrimination against the strikers. Later, all employees, union and non-union, would vote for a committee to act as their sole representative. On January 3, 1940, a notice was posted giving Dr. James' conditions for reopening the mine. They were probably much as Winch had anticipated, except that the company proposed to employ *equal* numbers of union and non-union men during an initial three-month period. Winch called it a "double-cross" since the union claimed to represent 65 percent of the employees at the time the strike was called. Viewed as attempted strikebreaking, James' proposal was rejected by a majority of the union members and the strike continued.[14]

Following this failure, Pioneer got serious about evictions. The task fell to Charlie Cunningham, a Bridge River resident involved in many things, including an appointment as a deputy sheriff. He agreed to act, but only on the condition that he be given a free hand in dealing with the 68 cases. At a ticklish meeting with the union men, he arranged that the evictions should be done at a rate of four men a day. At the same time he tried to find other accommodation in the local area for any who wished it.

Pioneer mine, striking miners on a sitdown, underground, February 1940. Photo courtesy of Irene Howard.

Cunningham handled matters with such tact and consideration that, strange to say, he won rather than lost friends among the miners.[15]

Growing desperate, the union men tried a new tactic. Early in the morning of February 27, 1940, some 41 of them broke into the Pioneer workings, descended the manway* in No. 3 Shaft, and began a sitdown strike in the hoistroom of No. 4 Shaft on 26 Level. This spot was probably chosen as it was dry and, with a constant temperature close to 65°F, reasonably comfortable for a long stay. Bud Rose of the mine staff and Sergeant Woods-Johnson of the Provincial Police descended to 26 Level and, on locating the strikers, agreed to leave the mine telephone and underground lighting in operation for the time being. Meanwhile, police reinforcements were rushed to the area.

The provincial government reacted the following day. Premier Duff Pattullo, declaring the strike illegal, ordered the strikers ejected from the mine and offered police protection, if necessary, to enable Pioneer to resume operations. Next morning, Inspector John Shirras of the police went underground to talk to the strikers He offered no prosecutions and his best efforts to get Pioneer to negotiate all issues if they came up peaceably but warned of the possible use of force if they refused. It was no idle threat; tear gas added to the mine ventilation system would have brought them to the surface in short order. The strikers came out that afternoon carrying the remains of their food supplies in their packs. In all they had been underground for just over 60 hours.[16]

The gamble failed. Pioneer management refused to negotiate and announced their intention to reopen the mine on March 6, 1940. Seventy-five men went to work that morning, ignoring boos from some 40 strikers clustered on the bridge over Cadwallader Creek and sullen stares from the 16 pickets permitted to stand in front of the dry.[17] So many police were on hand that the *Bridge River-Lillooet News* referred to a rash of "khaki spots" in the valley. Security was tight. The police kept an eye on everyone around the Pioneer camp, even insisting that moviegoers, bound for Bralorne, leave their names and check in on returning.

There were last-minute attempts to get the Bralorne miners involved in the dispute. On the evening of Thursday, March 7, Mine-Mill union representatives appeared in the Bralorne dry to ask support from the miners preparing to go underground. The co-operative asked for an hour to consider staging a 48-hour sympathy strike but abandoned the idea when the men simply walked out of the dry and went to work.[18] A government-supervised vote taken the same day confirmed the co-operative as the bargaining agent at Bralorne by a margin of 246 to 180.

On Sunday night, about 400 attended a meeting called by Mine-Mill's Local 271 at Bralorne to discuss aiding the strikers. It fizzled out when 300 walked out, led by a man who declared the meeting out of order since the co-operative was their bargaining agency and stated that, as a non-union man, he was leaving the hall and that other co-operative members should do the same.[19]

Forced to admit defeat, Pioneer's Mine-Mill local called off the strike as of 7:00 a.m. March 11, 1940. In all, they had held out for just over five months.

A few days later, the government offered transportation to those who wished to leave Pioneer, sending in a relief officer to issue the necessary tickets.[20] Not many union men stayed on, nor were they wanted. About seven were put to work on mine timbering, but others were stonewalled and told there would be nothing for them until the mine was in safe operating condition. Eventually most gave up and moved on to other mines, as did many of the non-union men, all convinced that Pioneer was no longer the place for them.

The strike left a bitter aftertaste and the Pioneer camp would never be quite the same again. Participants in the face-to-face confrontation, coached and prodded from behind the scenes, might well have been more tolerant of each other if left alone to settle their differences. As it was, the union movement, in too big a hurry for change, had lost a battle with the mine operators, but, given time, they would win the war. Looking back on it more than 20 years later, Sam Nomland, a member of Local 309's executive wrote:

> Though some critics have said that Mine-Mill lost the strike in 1939–40, I like to point out that there was, as a result of the strike, a very important point gained. The company, unwittingly, assisted in this regard by refusing to reinstate the most militant union men . . .
>
> These men, after the long and intense struggle of the strike, had gone through a school, you might say, and were now potential organizers . . . who scattered all over B.C. and waited in the various mining camps for the opportunity to reorganize.
>
> In all I believe the strike was well conducted and led. I, for one, would do the same thing over again-under the same circumstances. When all of B.C. organized (three years later) they found the only mistake was made by nature; the Pioneer strikers were only born three years too soon.[21]

Gold Bridge, 1934. Leonard Frank photo 17353, courtesy of the Jewish Historical Society, Vancouver.

# *Beyond the Mines*
## *1933–1941*

George and Margaret "Ma" Murray in 1967. Their *Bridge River-Lillooet News,* established in 1934 after an election promise, was a family operation for much of the time that the mines were active. Several owners later it is in its 66th year of publication. Photo courtesy of *Bridge River-Lillooet News.*

# *Settlements and Characters: Pioneer to Haylmore*

In the summer of 1936, about half the Bridge River's estimated 3,000 inhabitants lived outside the Bralorne and Pioneer camps.[1] Other settlements included both smaller mining operations and satellites to the company towns. There was more to the latter than beer parlours, bootleggers and, of course, the "sporting houses." Hotels were required to have a specified number of rooms before they could sell beer and, with the Bridge River crowded with men working on other properties or prospecting, all aspects of the hotel business were booming. And there were restaurants, garage and taxi services and many small retail stores, the latter often offering a different and possibly better selection than company stores. Gold Bridge was the largest community with the government liquor store, the highways yard and, about half a mile away, the mining recorder's office at Haylmore.

## *Pacific Eastern Junction*

This was the southernmost of the small independent settlements, located about half a mile from the Pioneer Mine. Better known as P.E. Junction or Shantytown, it was a cluster of buildings squeezed into the 66-foot road allowance. Somehow the highways department never seemed to get around to dealing with the trespass. Bralorne held the surrounding ground but some of the buildings had been there before that mine came into its own and, aside from huffing and puffing, there was little Bralorne could do except keep a close watch and block any further encroachments.

By mid-1935 the collection included Somerton's, a three-storey department store, plus three cafes, two taxi stands, a bath house, barbershop, shoemaker, broker and a dentist, together with several small houses. It was a firetrap, but it survived until late May 1939 when a fire, believed to have started from an overheated stove, destroyed most of the buildings.[2] There was no thought of rebuilding on the site, and Somerton's soon announced plans to rebuild at Ogden. Jim San, a Chinese merchant, had his new store, 35 by 60 feet, built on the Grull-Wihksne Mines ground just north of Ogden. For Dr. Sephton, a chiropractor, this was his fourth fire and, on losing everything, he abandoned his profession in favour of the insurance business.[3]

## Paddy the Shoemaker

Paddy, sometimes known as The Wayside Philosopher and one of the Bridge River's more colourful characters, made his home in a one-room, tarpapered shack in the road allowance midway between Pioneer and Bralorne. A man of tremendous girth, he spent his time reading, exchanging gossip with passers-by, carrying on a voluminous correspondence and, time and inclination permitting, repairing shoes. The shack was filled with books, many in piles on the floor with narrow pathways between, but despite the clutter, Paddy knew the location of each and every volume. He was interesting to talk to, reading giving him a vast general knowledge and local gossip a surprisingly accurate, up-to-the-minute account of goings on at nearby mines and prospects. Anyone prepared to listen would be treated to Paddy's opinions on the mining game, politics and the world in general.

The story of his early years, as revealed to one writer, may well have been larded with blarney to enhance his acquired image:

> Paddy the Shoemaker [Joseph P. Kieran, often spelled Kiernan], according to his own story, was born in Ireland, the son of well-to-do parents. He planned to become a doctor and had studied for two or three years when the Boer War interrupted his education. He joined the armed forces and when the war was over decided to remain in Africa.
>
> He became a Trade Commissioner in one of the African colonies and, because of his medical knowledge, gained considerable influence among the natives. In order to win their confidence, it was sometimes necessary to combine a certain amount of apparent witchcraft with the legitimate practice of medicine. On one occasion he treated the ulcer on the leg of the chieftain's daughter with a healing ointment. The treatment was accompanied by a potion of aniline dye taken internally. The resulting spectacular body waste matters won Paddy more renown than did the healing of the ulcer.
>
> After several years in Africa, Paddy returned to Ireland to work in his father's business. The arrangement proved unsatisfactory and Paddy came to America.

> After a short sojourn in the United States, during which time he developed an undying aversion for that country, Paddy came to Canada. One of his first investments was in a truck on which he carried scissors and knife-sharpening equipment. The Irishman began business on the eastern seaboard, and claimed that it was the most lucrative of any in which he had been involved. The only drawback was that a scissors grinder was required to obtain a licence. With this law, with true Irish obstinacy, Paddy refused to comply. He travelled across Canada plying his trade in each of various cities, until, in each, that wretched law was invoked, and Paddy was forced to move on to the next one.
>
> When he reached Vancouver he could go no further. He thereupon abandoned his trade and took a position teaching school in Lillooet. When it was discovered that Paddy had no teacher's certificate, he was given the choice of attending Normal School or changing professions. Red tape again![4]

From there it was on to the Bridge River and a new career as a shoemaker, his claim to fame helped along by whimsical items such as the following used to fill out the pages of *Bridge River-Lillooet News* in quiet weeks with little real news:

> Mr. Paddy Kieran, the wayside philosopher, who conducts a shoe repairing shop near Bralorne, had the following report to issue on his general state of being the other day:
>
> "I am physically alright, socially an outcast, politically damned, mentally uncertain and religiously a follower of Zoroaster, high priest of fire worshippers.
>
> "Bridge River was a good camp until university graduates, newspaper writers and stock salesmen damned the place.
>
> "Close all the schools for a year and revise the entire educational system"... University graduates, turned out with fur coats and cigarettes, like sausages out of a sausage mill, are unsuited to carry on affairs of the world and that is what put Canada and the United States in their present position.
>
> ...
>
> "I have given a night's lodging here at my shop to tramps who later made money in Bralorne or elsewhere and who now choke me with the dust of their motor cars as they dash by at reckless pace. I have fed hungry men who in a few months

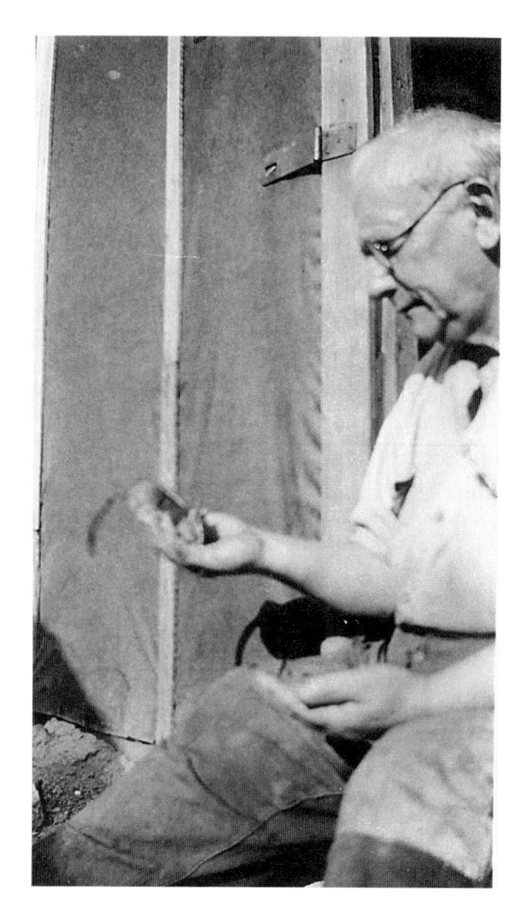

Paddy the Shoemaker (J.P. Kieran), 1934.
Photo courtesy of Mrs. D.M. King.

turned out to be socalled magnates. I am so puzzled by it that I give more and more time to half soling shoes and pegging away at my trade and refuse any longer to try to untangle the riddle of the Bridge River Valley and of the World for that matter.

"Some good women I know are thrifty and hard working, often making under-garments for themselves from the dainty flour sack. But their daughters demand and get silk hose and wear nothing but the finest silk next to their selfish skins. Though a lonely bachelor myself, my heart goes out to the heads of such families."

Quite well-known figures sometimes call at Paddy Kiernan's place. One day Ben Smith, the New York financier, dropped in.

> "Do you know me Paddy?" asked Ben Smith.
>
> "Know you; of course you are the lad that rustled the newspapers when me cousin was alderman of the blank Ward in New York."
>
> "The very same", said Ben Smith.
>
> "You are making a hell of a lot more money than when you were selling newspapers", said Paddy, "but you're not having half as much sport out of it."
>
> And so two Irishmen got together and had a long talk over old times.[5]

In another instance, news of a sort was offered mixed in with opinions and suggestions:

> If wayfarers along the road at Blackbird Point heard a loud fall last week, it was not rock tumbling down from the blast above Bradian, but it was Paddy Kiernan, the Wayside Philosopher falling out of his old rocker - Now Paddy is not ailing nor did he lose his balance, and it was two weeks after the holiday so he was absolutely sober, but the old chair creaked its last wail, crumbled up under the redoubtable Paddy and great was the fall there at.
>
> ...
>
> Being a cobbler is only a side line for Paddy Kiernan. It may be alright to sit on a packing box stool while he is pegging at boots and shoes.
>
> But if we are to receive his timely advice, his vehement criticism, his witty and infectious Irish humor, we must see that the chair is replaced.
>
> We suggest that friend Hutchings of Bralorne Bank be treasurer of an Endowment Fund for a Chair of Philosophy, which would lift our valley right into the Intelligentcia, and permit Professor Kiernan to continue with dignity and comfort his friendly advice to Bridge River Wayfarers.[6]

And so it went until late in 1939 when Paddy's seven-year career as a shoemaker was drawing to a close:

> There are always compensations if we just stop to possess them. Take Paddy Kiernan, the wayside philosopher of Blackbird

> Creek, who thought his world had come to an end when Father Time overtook him on a race and all but crippled him entirely. Bachelor Paddy lives in a shack on a sharp point between the two large Bridge River Gold Mines. When walking was good, Paddy used to repair shoes for a living. Paddy handed out caustic criticism and sound advice to the wayfarer who tarried. He was as happy as an Irishman can ever be. But when they took him to the Bralorne Hospital last week, his heart, he thought was broke entirely. Paddy would as soon be dead as around a bunch of women. He balked at the scrubbing brush. He winced at the shower bath, he fell out of the clean high bed. He mourned his old "dhudeen" (pipe). But he couldn't resist the smiling lovely nurses, the sweet sympathy and the devotion of his care, and Paddy is in love. At seventy, he has fallen like a ton of muck. He couldn't decide which one he loved the most so he said, "Sure I gave me heart to all o' them."[7]

Even the extra care wasn't enough and not quite a year later and after several stays in hospital, Paddy departed for the "banana belt" at Shalalth, leaving the Bridge River poorer for the loss. By now, Paddy had come up with his own scheme of beating the horse races, which he was willing to share for a mere $5. Ads in the *Los Angeles Times* and in other papers in areas where he thought there was more than the usual quota of suckers brought in a flood of money until the postal authorities caught up with him. Eventually, poverty and failing health landed him in the Old Folks Home in Kamloops where he died in late 1946, aged 77.[8]

## *Delina Noel*

A link to earlier days, Mrs. Noel still made her summer home at the cabin on the Lorne, a short climb above the Bralorne townsite. During the boom of the 1930s she was a local character known, or known of, throughout the Bridge River valley. She was alone now, the couple having separated around 1929, and Arthur Noel was now based at Vancouver's Castle Hotel.

Although not directly involved in the wheeling and dealing at the time of the Lorne consolidation, she had profited when Bralco made the final payment to Lorne Amalgamated Mines Limited. For their work on the Lorne property between 1916 and 1925, both the Noels had been issued shares representing a one-eighth interest in the latter company, and in a

Delina Noel late in life at her cabin.

series of payments each received close to $15,000 when the company was liquidated.[9]

Like many others in the boom she was much involved in the staking and selling of claims. A great talker, Mrs. Noel was full of tales of the early days and, in particular, of her version of a more recent dispute over the Noelton fraction on the northern boundary of Bralorne's Lorne group.

Mrs. Noel, acting as an agent, had staked the Noelton claim on October 10, 1923, and two weeks later the owner, Mary Eagleson, had turned it over to her for the customary "one dollar and other valuable consideration." In October 1928, Mrs. Noel had given an option on the fraction to J.M. Taylor, a Vancouver hardware merchant who was assembling blocks of claims around the main mining properties. The option calling for $10 down, $890 plus $100 towards the survey once the claim was Crown granted, and annual payments to bring the total to $5,000, with a final payment of $1,500 due October 1, 1931. Why the claim had to be Crown granted is not clear. Certainly it was not the usual practice, and if it was Taylor's idea he may well have outsmarted himself. In any case, the Crown grant was not issued until June 26, 1933, and presumably some, if not all, of the payments were still outstanding. Some two weeks before, Taylor had turned over his interest in the Noelton fraction to the newly formed Taylor (Bridge River) Mines Limited in return for 100,000 shares of that company. Taylor was trapped. He had to have the claim but his rights to it, if indeed he had any, were in dispute. The final price was high: an additional $65,000 paid to Mrs. Noel through a third party and a release, on behalf both Taylor and of Taylor (Bridge River) Mines, giving up all rights under the earlier 1928 agreement.[10] Ironically, the Noelton has gold showings, but they would not be discovered until the 1990s.

Mrs. Noel had a long-lasting love of a game of bridge. A mining engineer, writing of a 1924 visit to the Lorne mine, described meeting Mrs. Noel when she trammed an ore car out of the adit where Arthur was mining, and how nothing would do but that Dave Sloan, then the closest neighbor, be summoned from Pioneer to make up the foursome for a game that evening.[11] She was still at it and, on a summer evening, her guests from Bralorne would walk up the hill for a game. For just as long she had been opposed to liquor and all its evils but, on special occasions, her guests might be offered a single drink and, at the end of the evening, served hot Ovaltine and cookies to fortify them for their walk back to Bralorne.

## Ogden

By March 1934, Ogden was advertising itself in the *Bridge River-Lillooet News* as "50 feet from the Bralorne Mine - The Payroll Town." That was stretching things; the "50 feet" was from the Bralorne property line, the camp and portal were some half mile away, and not much of the payroll could be tapped until there were more places to spend money. Location aside, the site was a poor one, with buildings perched precariously on the hillside both above and below the road. Snow and mudslides were a menace during spring thaws, and from time to time the highways department would threaten to remove any buildings trespassing on the road allowance, an act that would "put half of Ogden in Cadwallader Canyon."[12] Still, if a building site could be found or created, Ogden was the obvious place for a hotel and accompanying beer parlour.

E.P. Brennan undertook the hotel project. It is difficult to understand why; Brennan knew very little about the Bridge River and lacked experience in the hotel business. The general manager of a pulp mill at Woodfibre, near Vancouver, he may have been sold on the opportunity by Austin Taylor or one of the other Bralorne directors.

Burroughs and Wright, who operated a general store, owned the necessary land, and a suitable flat spot could be created if the government could be persuaded to shift the road allowance onto fill dumped over the bank. Brennan went right to the top, approaching Duff Pattullo, the premier of B.C., about it. Pattullo told him that something might be arranged but that it would have to be done through a senior officer in the Lands Department. Unfortunately for Brennan, he appeared in the latter's office before word of the proposal filtered down from above. Taken by surprise, the official, a very proper Englishman, peered down his nose at Brennan and told him, "You cawn't build on His Majesty's Land." Brennan, trying again to explain and getting nowhere, let fly with his Irish temper, damning both the official and His Majesty, King George V.

Storming back to Pattullo's office, Brennan was told that nothing could be done for him unless he went back and made a full apology to the self-important Englishman. Brennan swallowed his pride, did it and, in due course, was granted the dispensation. Years later it still rankled, Brennan claiming the apology one of the most difficult things he had ever had to do.[13]

In July 1934 a crew using a gas shovel started work on the excavation for the Mines Hotel. It took months to complete. In all, some 22,000 cubic yards were gouged from the hillside to create a flat spot at road level large enough for the 80- by 36-foot foundation.[14] Local highways officials, unhappy about it all, made things difficult by refusing permission to close the highway, even for short periods, and insisting that the full road allowance be left in front of the hotel site. The hotel, an impressive wooden structure with two dormers and a massive stone chimney inset near the centre of its front face, opened for business on December 20, 1934, with Brennan's brother, Robert, as manager.

It lasted less than a year, going up like a box of matches on Monday morning, October 7, 1935.[15] Fortunately, there was no one in the building at the time. Despite the promising location, the hotel did not appear to be doing well, giving rise to speculation over the origin of the fire and the thick black smoke that poured from the burning building. It was rebuilt in the spring of 1936, using the same plans. The reincarnation lasted a little longer, burning on Monday morning, May 17, 1937. This time the fire

The original Mines Hotel about the time of opening on December 20, 1934. The building was destroyed by fire on October 7, 1935. Photo by W.A. Hutchings.

started in the kitchen and those inside managed to scramble out before the building was engulfed.

Until the hotel was rebuilt, thirsty miners had to take their business down the road to hotels in Brexton and Gold Bridge. The *Bridge River-Lillooet News,* based in Gold Bridge at the time, reported that "even the News woodshed has been pressed into service by providing sleeping quarters for some of the consumers of the foaming suds who found themselves stranded ten miles from home."[16] This time, fancy plans were tossed aside and the Mines Hotel that rose from the ashes was a boxy, utilitarian structure that would last until its turn came in the summer of 1984 and the unoccupied building burned to the ground.

For Brennan, the hotel venture had turned into a nightmare. After rebuilding twice and with poor management, he was close to $90,000 in debt to a Vancouver bank and had taken to crossing to the other side of the street when passing it. Forgetting one day, he was hailed by a clerk and told that the manager wanted to see him. Instead of the confrontation Brennan was dreading the manager wanted to know if Brennan, whose wife was seriously ill at the time, needed money to meet the emergency. Brennan demurred, saying that he saw no prospect of paying back what he already owed. To his surprise, the manager said that had nothing to do with it and insisted on advancing another thousand dollars to be paid back when Brennan was able.[17]

Things changed in October 1940 when John Branca, part-owner of the Royal Hotel in Gold Bridge, took over management of the Mines Hotel on a two-year contract. At the time the hotel had a poor reputation among the miners, and at one point there had been a campaign to blacklist it for serving short beers.

Branca knew the hotel business, and in short order he was on good terms with both miners and mining companies. One policy differed from that of most hotels: no parcel went into the safe unless Branca was permitted to examine the contents. Anyone who refused to co-operate was invited to take his parcel elsewhere, and some did just that, stamping out of the hotel in disgust. Branca wanted to be certain that no one could accuse him of aiding or abetting high-graders rumoured to be active at both mines.

The greatest change was in "The Big Stope," the hotel's beer parlour. Out went the pitcher under the taps, used to catch slopped-over beer and add it to someone else's glass. No more short beers either. From now on

The third Mines Hotel about 1950, with John Branca in the foreground. The building opened early in 1938 and was destroyed by fire on June 7, 1984. Photo courtesy of P. Branca.

the beer in each glass, not the foaming head, must be at or above the 8-ounce "Plimsoll line" enameled on government-approved beer glasses. There was another side too: Branca kept a close check on the "draw" that each of his tapmen got from a 24 gallon keg of draft beer. A poor draw could drop the revenue from each barrel by $6 or $7, something over 10 percent. With the changes, Branca won the goodwill of the miners and The Big Stope was often filled to its 250-odd capacity.

For the most part, the customers were an easy-going, happy crowd. Troublemakers were given one warning and then refused service if they acted up again. Underage drinkers were called into Branca's office, asked to sign an affidavit that they were 21 years of age, and warned of the consequences of a false oath. That usually discouraged them, although it was all bluff with the government form no longer in use. Mining was the main topic of conversation, with everyone boasting of their great prowess. Bralorne officials took it all with a grain of salt; if their mine produced at the boasted rates it would have been a world-beater!

On occasion the drinkers would burst into song, something forbidden under the liquor laws then in force. Police from the local detachment ignored it but, as luck would have it, the crowd was in full voice on a night when a visiting officer and sergeant of the Provincial Police checked into

Phil Branca (right) and brother Dick behind the bar of the Mines Hotel "Big Stope," 1953. Photo courtesy of P. Branca.

the hotel. That did it; a few days later word was out that the local detachment had been ordered to enforce the rule. The sergeant and a constable appeared in The Big Stope and the former stood staring at the licence posted on the wall until the room quieted down. Commenting, "Beer - no singing; Singing - no beer" he turned on his heel and walked out. Branca dimmed the lights and went around to the tables explaining what had happened. A few hotheads, anxious to challenge the ruling, were invited outside by fellow patrons and convinced, by numbers if nothing else, that it would be in their best interest to keep quiet.

Thirsty patrons were often permitted to cash postdated personal cheques. The risk was minimal; through an informal but mutually beneficial arrangement with bankers at both mines, Branca had credit limits for many of his customers. For a single miner it could be as high as $30, while a married man was cut off at a flat $2, unless his wife was sitting at the table and nodded agreement to a larger amount. The cheques, mostly small, were bundled off to the banks whose total charge of 50¢ for holding and cashing them made a nice sweetener to branch accounts. To return the favour, if one of the banks ran short of bills or coins, a phone call to Branca would see the required amount, sometimes as much as $3,000, packed into a canvas bag and dispatched by taxi.

Branca had a long and happy association with the Mines Hotel. Beginning with the first contract in October 1940 he stayed on, buying out Brennan's interest in 1948 and operating the hotel until 1955 when he sold out and left the Bridge River. He was the last of the original hotel men; most were gone by the early 1940s. The Mines Hotel continued to operate under a number of owners until the 1984 fire. It was the last to go; other buildings that once clustered around the site had disappeared years before.

Ogden's other amenities included a general store, restaurants, dress shop, dentist's office, combined taxi stand and smoke shop plus a small settlement above the road. The "smoke shop" was actually a cover for a gambling room, but the gamblers kept to themselves, turning away anyone they didn't recognize, and the police, for their part, simply looked the other way.

Another landmark was Zada's sporting house, once described as "a well-run and highly respectable 'disorderly' house."[18] Its claims to fame included quick repairs and return to "business as usual" after a slide of snow and mud landed in the middle of the building. Zada Fontaine herself was a tall, white-haired woman, stately enough to turn heads when she walked her dog on Vancouver's Georgia Street but, according to some, flamboyant enough to send her dog over to the Ogden barbershop for a trim. Rumour had her associated with some well-connected people, including at least one provincial politician. She was kind to her girls, most of them in or just out of their teens, and was always willing to do what she could to help out a would-be miner who had gone broke while hanging around the mines in hopes of landing a job.

Above all, Zada knew how to handle things. When a group of Bralorne youths, curious to know what went on, pooled their money to send the one with the short straw up the steep steps to her establishment, the would-be customer came flying down so fast that he missed most of the stairs along the way. After Zada left, her replacement, lacking class, handled things differently. One of her girls, wearing only panties and a bra, was badly scraped and scratched after a wild dash down the hillside a step ahead of the madam's waving butcher knife.

## Brexton

Known as Fish Lake until 1938, Brexton grew on lots leased from B.R.X. Consolidated Mines located at the edge of a low, swampy area where the Bralorne road swings close to Fergusson Creek. The location had two strikes against it: Ogden was much closer to the mines and potential customers, and Gold Bridge had a much better site with ample room for

Brexton (Fish Lake) about 1935. Photo by W.A. Hutchings.

The Truax Inn at Brexton, about 1934. Photo by W.A. Hutchings.

growth. Still, by August 1934, it boasted the Truax Inn, complete with beer parlour and the nearest oasis when the Mines Hotel was being rebuilt after the two fires. Other businesses added during the boom included a tailor shop, jewelry store, combined butcher and grocery store, cafe, garage and perhaps a dozen private homes and a school. Brexton declined slowly, but the Truax Inn managed to keep operating until some time after World War II, when it was cut in two and moved to Gold Bridge to replace a hotel lost in a fire there.

## *Gold Bridge*

Gold Bridge's birth in 1933 was described in the first issue of the *Bridge River-Lillooet News* datelined March 1, 1934. Like the town itself, the paper was an offspring of the Bridge River rush, an election promise made by George Murray, winner of the 1933 contest to represent the Lillooet riding in the Provincial Legislature:

> The promoters of the modern Townsite of Gold Bridge, chose a site on the banks of the Bridge River at the foot of the hill leading to [the mines and] commanding a lovely view of the

wooded banks of the River, both East and West. The Townsite was surveyed and lots staked under the supervision of J. Harris & Company Limited, of Vancouver, the streets being laid out and graded. An abundant supply of good water was obtained from a creek a distance up the mountain, and a three inch wood stave main laid along the whole mile, at a considerable cost to serve the principal section of the Townsite. A few months ago there only existed a trail at the East end, where the main Government Highway passes up the hill a few shacks and cabins along the route.

The inspiring sound of hammer and saw were soon heard in the erection of a modern, up to date, fifteen room hotel, The Gold-Bridge Hotel. The Government in its wisdom decided that a glass of beer would not be harmful to the lusty miner, so granted a license.... There have quickly followed the usual stores in a town which holds no parallel to the "Boom Town" of pungent memory.

The B.R.X. camp in 1934. This property between Gold Bridge and Bralorne was one of many that never quite made it. Despite serious exploration beginning in 1896 and almost 20,000 feet of underground workings, only a small amount of gold and silver were produced. Photo by W.A. Hutchings.

Next to come was a laundry built by Dan Eng with the help of a loan from Charlie Hodgins, the hotel keeper. Before long, Eng had six others working for him and a corner on the laundry business, aside from the Pioneer camp, which had its own. Stores, garages, cafes, a lumberyard, and a branch of the Canadian Bank of Commerce followed almost as fast as the buildings could be slapped up. The sporting houses were there too, just outside the town limits. The first grocery store, built by Al and Neil Trout, outgrew the original building in six months and had to be rebuilt. The boom was on: Pioneer and Bralorne were doing better all the time and, with similar-looking quartz veins being explored all through the area, there could be dozens more mines like them.

Typical newcomers were Charlie and Evelyn Cunningham, who arrived in the spring of 1934 with plans to set up an insurance business. Charlie, born in Ontario, had once had hopes of getting into professional sports but, when that failed to work out, had drifted west doing a number of things, most recently working as an electrician in Consolidated's big Sullivan mine at Kimberley. Now, determined to be on their own, the couple had reached Gold Bridge.

Their lot, bought on Charlie's scouting trip the year before, was across the street from the Goldbridge Hotel. Working with the help of an older man, the Cunninghams had their building, with front office, living room, bedroom and kitchen, roughed in after two weeks. Fortunately the old-timer let them use his cabin until their own place was ready. The latter was small, with just enough room for a tiny stove and a rough bed covered with cedar boughs. But, most important of all, it gave the couple a little privacy in the midst of all the surrounding activity. By July the insurance business was getting established and the *Bridge River-Lillooet News* now carried a tiny ad: "Cunningham at Gold Bridge pays good money for ashes."

When winter came the Cunninghams found their new home next to impossible to keep warm. The only heat came from a small kitchen stove with a drum oven in the stovepipe and, despite constant stoking, the rooms were seldom comfortable. In the spring they were told that part of the building lay on the road allowance, and on rebuilding on a sloping portion of their lot, a downstairs room was added in back. Rented out, they found their tenant contributed not only cash but also heat from a stove, a big help in keeping the upstairs warm.

The Cunninghams stayed on in the Bridge River except for a few short absences in later years. Charlie tried his hand at and was successful in any number of occupations, including big game guide, notary public, deputy sheriff, chicken rancher, wildlife photographer, insurance agent,

Charlie Cunningham on frozen Gun Lake in the early 1940s. Photo courtesy of B. Orgnacco.

electrician, real estate agent; so many in fact that a listing in Will Haylmore's flowing script came close to filling the front of an envelope addressed to him.

Gold Bridge kept growing and by December 1934 boasted some 26 businesses, three cafes and two hotels, some of them not quite completed, two general stores, two barber shops, bank, newspaper office, clothing, bakery, drug and furniture stores, pool and tea rooms, lumberyard, laundry, cleaner, taxi and wrecker service, plumbing and sheet metal works, accountant, insurance agent, radio service, and a library and magazine exchange.[19]

Government was there too, represented by the post office, a telegraph office, the highways crew, and a mining recorder's office at Haylmore about half a mile away. A government liquor store opened just in time for the Christmas rush replete with 13 tons of stock. It wasn't enough and the shelves were empty before the holiday began.

The school, a new building 24 by 32 feet, opened on schedule in early September with 13 pupils taught by Miss Ruby McAllister. The playground, covering nine government lots, still had not been cleared and the building lacked cloak rooms, inside plumbing and a library, all estimated to cost an additional $2,000. The school board, perhaps feeling that they

The Royal Hotel at Gold Bridge, 1935. Photo by W.A. Hutchings.

had done enough for one year, decided to pay off the outstanding bills and put the rest off until next year.

During 1934, the business men organized their Board of Trade and the women their Ladies Gold Nugget Club. Both had their own special interests but inevitably they were drawn into the controversy over three sporting houses that had sprung up on the edge of town. With hundreds of men in the bunkhouses at the mines, there was no shortage of patrons. These arrived at all hours in varying states of intoxication, often announced by squealing tires and clouds of dust. The Board of Trade was in favour of the houses since they brought money to the town's other enterprises. The Gold Nugget Club split on the issue, some members considering the houses a necessary evil, but others insisted that they should either be closed or forced to move elsewhere. The school board, concerned about the dangers the cars posed for the children, let alone any moral considerations, wanted to be rid of the houses.

The *Bridge River-Lillooet News* favoured a fair shake for the sporting-house girls and Ma Murray took on the opposition in a hard-hitting editorial:

> They were here first.... If you want them to move, let's get in

In addition to Bralorne and Pioneer, several other mines operated briefly in the Bridge River valley and produced small amounts of gold and silver. One of these was the Wayside property near Gold Bridge, here seen in a 1934 photo by W.A. Hutchings.

> and help them move. Allow them uptown on mail days to shop. Stop harrassing them. After all, Gold Bridge was happy enough to sell them the lots and lumber for their houses. They don't bother anybody. And it's fine to adjust your rose-coloured glasses and say these houses shouldn't exist. But until the six hundred single men in this valley can build themselves homes and bring their wives and families here, you may as well relax and face the fact that sporting houses are necessary to the safety of the women who do live here.[20]

The policy of live and let live had its moments of humour, such as one mail day at the store when the priest asked Alice, a real beauty, if he hadn't met her somewhere before. There was pathos too, in the hypocrisy of a code that saw one of the girls pass a schoolgirl almost her own age on the street without any sign of recognition, this a few days after a stay in the family home to nurse the mother through a serious illness.

Gold Bridge prospered for a few years, then slowed down as the exploration boom fizzled out on failing to find new mines to match Bralorne and Pioneer. The wheelers and dealers and the fast buck artists moved on and, more and more, Gold Bridge's fortunes waxed and waned with those of the two big mines. It was still a lively place with more than its share of characters, most harmless, but a few with a glint of larceny in their eyes, plus a growing number of families who now considered it their home.

## *Haylmore*

The office and home of Will Haylmore, sub-mining recorder for the area, lay half a mile from Gold Bridge on a low bench overlooking the Hurley River. An old-timer, he had staked the Why Not claim in June 1897, and since then he had been involved in many ventures in the camp, none of them particularly successful. His stories of the early days, plus landscaping that made his place "a garden of Eden compared to the rest of the Valley," put a courtesy call at Haylmore high on the agenda for visiting dignitaries.[21]

A tall, spare man with white hair, Will Haylmore had a trademark of sorts in a white silk scarf worn about his neck on all but the most formal occasions. That, plus his erect carriage, gave him a touch of class evident even when heading for the bush in pants with the knees out and with long underwear, the old-time prospector's standby both winter and summer, poking out from under his shirt cuffs. He took himself seriously and

Will Haylmore, with one of his many rock walls, about 1935. Note the machine gun behind his head and shoulder. Photo by W.A. Hutchings.

expected others to do the same. Once this cost him a lunch when, invited to the home of the Gold Bridge hairdresser, he stormed out after a youngster looked up at his flowing white hair and asked if he had come for a haircut or a perm.[22]

Accounts of his early years varied, quite possibly because Haylmore intended they should. All began with his birth in England, but one visitor who knew him well was told he had run away from home at fifteen, joined the navy and after a few years drifted to America. Others were told or given to understand that he was a remittance man, disowned by his family and shipped out of England. In any case he remained a royalist; the Union Jack flew from the flagpole outside his office and the interior was decorated with pictures of the royal family clipped from *The Illustrated London News.* At some point in his travels he had been bitten by the prospecting bug, and there were tales of work in California and the Cariboo before arriving in Lillooet in the 1890s. He stuck to prospecting, even after marrying Maude Manson of Lillooet in December 1909, and, for the first years together, the couple combed the hillsides above Seton and Anderson lakes. Then, in 1916, when in charge of work on the Coronation Mines property, later part of Bralorne's ground, he had discovered and trenched a gold-bearing quartz vein parallel to the main Coronation Vein.[23]

At some time in the 1920s Haylmore went back to placer mining, acquiring both a half-mile creek lease and a bench lease on the Hurley River near his home. Rich gravels had already been worked on the ground and even better gravels were believed to lie in the bed of the Hurley River. Three previous attempts, the first in the 1880s, had been made to divert the river into a new channel against the west, or opposite, bank by means of a wingdam,* but all had failed when the new channel, unable to carry the flow, had plugged with silt and debris, raising the water level until it

WIll Haylmore's home, which also served as the mining recorder's office, about 1935. Photo by W.A. Hutchings.

topped and destroyed the dam. Haylmore, in less of a hurry, hoped to prepare an adequate channel before diverting the river. For years, small crews, mostly unemployed men given a few days work, had built small dams to turn a portion of the flow into the new channel in an attempt to deepen it.

Despite official duties as mining recorder and the placer work underway on his leases, Haylmore was still very much interested in lode deposits. He grubstaked prospectors and, for a time in 1935, there were reports that he was teaming up with Ben Smith, Pioneer's master promoter, to develop a quartz vein in the Blackwater area near D'Arcy.[24] In addition, he was caught up in and perhaps the source of some of the excitement and rumour that always swirls around a booming mining camp. All this bothered some prospectors, who felt that Haylmore, tipped off by the documents flowing through his office, might have an unfair advantage in spotting and profiting from any new staking rush. They preferred to travel to Lillooet to record their claims, counting on a brief delay before news of what they were up to got back to the Bridge River.

As mining recorder, Haylmore ran his office his own way. Even when swamped with work at the height of the boom, he made entries and copied documents longhand, refusing to use carbon paper or have a typewriter on the place. At times prospectors were lined up at the door before the office opened in the morning and, with no time for cooking, Haylmore and his assistants took turns going into Gold Bridge for a restaurant meal.[25] He had a heart too; if a prospector was unable to pay the recording fees, Haylmore would pay it out of his own pocket in return for work on his rock walls or other projects.

Visitors to Haylmore's cabin often brought a gift of Scotch whiskey or Geneva gin, two favourites, and, in turn, Haylmore would serve them drinks with the "consummate hand of a born actor - beautiful glasses linen cloth and a flourish."[26] Then, if the mood was right, he would regale his guests with tales of early days in the Bridge River. Some involved Hunter Jack, the Indian who, in the 1880s, had chased Chinese miners out of the valley and, according to Haylmore, had filled a five-pound baking powder can with gold nuggets taken from a small area of bedrock located 100 feet or so from where they were now sitting.[27] Another was of attending a potlatch given by Hunter Jack at which the latter distributed the contents of a two-quart pail filled with nuggets to a group of native women seated in a circle.[28] Other yarns were of Haylmore's own operations and escapades, not always to his credit, but full of the joy of life and the fun of outwitting would-be rivals.[29]

Listeners may well have suspected that Haylmore's stories grew a little with each telling, but no one questioned his generosity to the job-seekers who flocked to the Bridge River. Forestry crews, put to work on projects he had organized and got government money for, were sometimes kept on after the funds had run out and paid from his own pocket. In one reporter's words:

> No man in search of honest work ever went away from Haylmore with an empty stomach. In fact, it is Mr. Haylmore's policy to start each year flat broke and give away what he makes as he goes along. He is an integral part of Bridge River - in fact, an institution.[30]

True or very close to it, and a fitting tribute to the man.

Bralorne residents off to a holiday celebration in Gold Bridge with driver Louis Santini (left). Possibly May 25, 1936. Photo courtesy of M. Jukich.

# *The Road to the Mines*

With Bralorne and Pioneer prospering, large tonnages of freight had to be hauled from the railhead at Shalalth, and a direct road link to the mines was still many years in the future. Both mines left freighting to independent transportation companies, with much of Bralorne's business handled by the Neal Evans Transportation Company and Pioneer's by a smaller rival, Bridge River Motor Services. In addition to their work for the mines, both companies carried freight and passengers for others and acted as distributors for rival oil companies. There was other competition too: for freight from smaller operators, both licensed and scofflaw, and for passengers from a number of taxi operators.

In 1934 the Evans fleet consisted of seven trucks and four Cadillac cars for passenger service while the Bridge River Motor Service had four trucks.[1] By 1940, its best year, the Evans fleet had grown to ten trucks of various makes, four Chrysler Airflow stages and a 29-passenger bus, and the company offered twice-weekly bus service from the mines to Vancouver, with the bus moved by rail across the gap between Shalalth and Lillooet.[2]

Neal "Curly" Evans, owner of the company bearing his name, was one of the many colourful characters in the Bridge River. Born in the Dakotas in the late 1880s, he was in Ashcroft in 1910 driving jerkline teams of horses hauling freight wagons on the Cariboo Road. His present company had grown from a taxi service started in 1927 from Shalalth with Louis Santini as the first driver. By 1934, in addition to his trucks and stages, he was considering offering air service to the Bridge River. Already Curly and "Ginger" Coote, a well-known pilot, were involved in a separate company operating a four-passenger Fairchild aircraft and, as part of that deal, Coote had given him flying lessons.[3] But scheduled float plane service between Gun Lake and Vancouver never materialized despite much talk and a brief attempt to fly the route in 1939.

Curly never lost his love of horses and continued to dress as a cowboy, with high-heeled boots, pants slashed off at the bottom so they could be pulled on over the boots and a soft, black felt hat with the brim turned up all around. Some winters found him taking in the races at the Santa Anita track and, closer to home, he revelled in the Minto stampede put on by his buddy Big Bill Davidson. He was a bit of a cowboy with his trucks too:

Last week on the Bridge River road one of the big Hayes Anderson trucks... slid off the bank... turned over twice and dived into the muddy slime of the river.

The driver was knocked unconscious and smashed up. The truck... settled comfortably into the mud until the great cables dragged it up the bank. Sitting in the middle of the road its wheels on terra firma again, Curley stepped on the gas, it heaved and coughed, mud dripping from its innards and water

Neal "Curly" Evans. Photo courtesy of Lillooet Museum.

flew from the chassis, he put the clutch in low and released her, and she clumsily started out, and came into Shalalth over the mountains on her own steam, "All in a day," stated "Curley".[4]

One innovation directly related to Curly's cowboy days was the development of ice hooks to give better traction under extremely slippery road conditions. Horseshoe caulks, conical steel studs about an inch long designed to be threaded into holes in a horseshoe when the animal was working on ice, were threaded into specially curved steel bars mounted across the face of an ordinary tire chain. Simple but effective, they filled the bill.[5]

By late 1933 the provincial government was claiming to have provided a "good auto-road" to the Bridge River that had seen 10,000 tons of freight

The telephone at 7 Mile on the Bridge River side of Mission Pass. Drivers of Shalalth-bound vehicles were expected to call ahead and confirm that the road was clear before starting their descent. Photo courtesy of Mrs. D.M. King.

trucked over it that year and had "every accommodation in the way of hotels, gas stations, restaurants etc."[6] Few who drove it would agree; possibly the government was talking to itself, spreading the idea that there was no need to spend more money on improvements. In any case, it was still a narrow, mountain road without guard rails, lightly gravelled, full of hairpin turns, and subject to the vagaries of mountain weather that sometimes closed it in a matter of hours.

There were many bad spots along the road, but the worst section was the five miles of switchbacks on the steep, 3,000-foot climb from Shalalth to the summit of Mission Pass. In winter, the open south-facing slope was swept by snowslides, often catching deer feeding there and, with a thaw, the appearance of their mangled remains was a grisly reminder of just how treacherous it was. At other times, a quick thaw followed by freezing weather could coat the road with a sheet of ice. In summer, extreme heat was a problem and it was best to start the climb about 4:00 in the morning in hopes of reaching the pass before the radiator boiled over. Ascending traffic had the right of way and drivers heading to Shalalth were expected to use a phone at 7 mile to call ahead and confirm that the road was clear before starting their descent.

The *Bridge River-Lillooet News* of April 21, 1938, described a miraculous escape on this section of the road:

> One of the most amazing tests of the modern motor car from a safety standpoint was staged on Monday on the Bridge River highway when a 1937 Ford, driven by Harold Wales, Dubois Motors, Goldbridge, went over the side on Five Mile hill out of Shalalth.
>
> The Ford turned two head-on somersaults, then hit a tree, rolled over on its side, jumped into a ravine 200 feet below.
>
> One man, two women and two children were in the car with Driver Wales. No one was seriously hurt. In fact, no one went to hospital.
>
> The car itself was a tangle of wreckage. All-steel top and safety glass had proved their worth under the most strenuous test ever given in the Lillooet mountains.
>
> So striking was this demonstration of the achievement of the modern motorcar manufacturer that lifelong boosters for rival makes in the persons of drivers on the Bridge River highway for the transportation companies joined in paying tribute to the car which proved such a protector of life and limb.

> Wales, known as a first-class driver, cannot explain exactly what happened to cause the car to leave the road on the most dangerous point between Shalalth and Pioneer.

On another occasion an Evans driver, aware that his brakes were slipping but determined to get his passengers to Shalalth in time for the train connection, tackled the problem by cutting down a good-sized tree, tying it behind the stage and dragging it down the Mission Mountain switchbacks. It worked: he got down safely and in time and, in the process, added a new twist to the Mission Mountain saga.[7]

Little wonder that some residents of Shalalth and Seton Portage, aware of what went on, elected to stay right where they were, flatly refusing to risk their lives on a trip to the Bridge River.

Transport drivers took great pride in getting their loads through despite weather and road conditions and the awkward nature of some heavy loads of mining equipment. To accomplish it they had to be ready to cope with

Neal Evans Transportation Company trucks on the Mission Mountain road. Photo courtesy of Neal Evans.

an emergency at any moment. When things went wrong there was seldom any outside help to call on, and those at the scene pitched in to get things rolling again. Of course, there were breakdowns and accidents. Some of the latter were unavoidable, such as being caught in a slide or having the shoulder of the road give way, and a few inexplicable except as momentary lapses:

> ... Louis Santini, whose 10 year record of driving cars on the Bridge River roads, unsmirched, was knocked into the Cadwallader Canyon Wednesday a.m. and nearly with it two passengers ... when Louis left his Evan's Chrysler stage at the Mines hotel door to scurry in to oblige the collection of way passengers. He left the car in high gear with the brakes on, with passengers in the back seat. The car started rolling and as the pair knew nothing about stopping it, they scrambled out. The car took a nose dive over the precipitous bank, and stopped only at the bottom, 350 feet below. Evans trucks and lines hauled it back up, dinted and bruised, but not completely wrecked. The lava ash is loose there anyway.[8]

Perhaps the greatest challenge faced by the road and transportation company crews was the cleanup in the wake of a monster snowstorm that struck in late January 1935. On January 19, the road was blocked by a slide near Falls (now Fell) Creek, about seven miles upstream from the first crossing of the Bridge River. A stage driver who saw the slide come down just behind his Packard car waited until the drivers of two trucks that were following him made their way across the slide on foot, and then the trio drove on to Shalalth to get help.

Next morning a seven-man crew with two trucks was dispatched to clear the slide but, delayed by new slides, they arrived after dark. Unable to assess the damage to the road, they put up overnight at a nearby camp. By morning, snow falling at the rate of four inches per hour had covered their trucks and new, larger slides now blocked the road behind them. Confident that a plough would get through within a day they decided to stick it out, but on January 22 with food getting low two of the men set out on foot for the nearest telephone at a hydro camp near the bridge. They made it, but only after 12 hours struggling through snow up to their waists and, in two places, crawling across slides that might let go at any moment and sweep them into the river below. Exhausted and too stiff to move for an entire day, the pair spent the next few days recuperating.

Meanwhile, the government bulldozer based at Shalalth had cleared to the pass on Mission Mountain only to have to retrace its route when new slides closed the road behind it. A relief expedition was organized by Frank Chapman, general manager of the Evans company, using that company's small 30-hp bulldozer to tow a bobsled loaded with provisions, fuel and oil. Dubbed "South Pole Expedition No. 2," it got through to the hydro camp. From there two Indian packers, Billie and Henry Patrick, set out for Falls Creek carrying food and snowshoes. They made it and the entire party snowshoed back to the hydro camp and then rode the bobsled back to Shalalth, arriving tired and hungry on the afternoon of January 25, five days after the first party had set out.

Four 60-hp bulldozers, two each from the mines and Shalalth, were used to clear the road through drifts and slides as much as 21 feet deep. Another crew worked night and day to save the bridge near Gold Bridge, using dynamite to blast the ice jams piling up against it. It was a close call and at one point backed-up water was lapping over the road. The road crews working from each end met at Falls Creek at 5:00 a.m. on January 30, and the first truck pulled into Gold Bridge at 4:00 p.m. the same day after making the run from Shalalth in three hours flat.[9]

Aside from the struggle to clear the road, two men lost their lives in a snowslide at Ogden. Another slide blasted through the Bralorne cookhouse, and both mines suffered severe power cutbacks and came perilously close to losing their power plants completely.

Constable R.S. Welsman and the B.C. Provincial Police station at Bralorne, 1935.
Photo by W.A. Hutchings.

# *Policing the Bridge River*

By and large the Bridge River was law-abiding. The British Columbia Provincial Police, covering it with a two-man detachment based at Bralorne, could not be everywhere and followed an informal policy of "live and let live" that ignored prostitution and gambling and instead concentrated on accidents along the road, trouble-making drunks and, infrequently, more serious crimes.

Prostitution was never legal but, with so many single men around, general opinion held that there would be more problems if the sporting houses were closed down. Considering the interest they still evoke, the number of houses was quite small. Beginning with Zada's in Ogden, the list includes the Canyon Cat along the road towards Gold Bridge, Ma Kimble's and two others in the "Hollywood district" just outside Gold Bridge, "Old Mother Larders Arse" near South Fork at the mouth of Hurley River, and Helene's in Minto. Prices were more or less standard, 50¢ for a drink, $3 for a trick and $15 for an all-night stand. Often a "hi-pillow boy" worked for the madam, cutting wood and hauling water in return for food and lodging and possibly a little spending money and other unspecified fringe benefits. Just what, if anything, the latter amounted to is still speculated about by those who never held the job.

The madams kept a tight control on their operations, dealing with drunken miners, settling disputes among the girls and, most important of all, ensuring that no one, whether a customer or simply a neighbor, had grounds for a complaint to the police. One afternoon each week the girls were permitted to visit the company towns to do their shopping and banking. It was a big event when the madam and her girls, all dressed in their finery, wheeled into town by taxi. Aside from this outing they were next to invisible, banned from all the social events in the community.

The police had little serious crime to contend with, in part because of the isolation, as there was a good chance of stopping anyone on the run by phoning ahead to Shalalth and Lillooet. Despite the odds, the Bank of Toronto branch at Pioneer was robbed twice, in August 1935 and again almost seven years later. As the *Bridge River-Lillooet News* reported the first:

> The Pioneer bank robbery took place on Tuesday morning [August 20, 1935] when two bandits stuck a gun into the face of

> J.E. Boyle, manager of the Bank of Toronto, who lifted his head from a ledger to look into the barrel of a big revolver. The first of its kind in the valley, the robbery has kept local residents in a constant fever of excitement, more perhaps than might have been brought about by a rich strike or a disaster.
>
> "Stick 'em up," said the biggest.
>
> Banker Pat did not even bother to lift his head from the column of figures he was checking.
>
> "Stick 'em up, I tell you. This is a holdup."
>
> "The Hell you say," smiled Pat and he raised his eyes to look square into a gun, held by the palsied hand of an amateur bad man who was unmasked and quaking with fear. Banker Boyle was not long in throwing up his hands.
>
> "As a matter of fact," Pat said "I acted quickly for fear those shaking fingers would hit the trigger before I got my hands high enough!"

Boyle and Mrs. Warren, a customer who came in during the robbery, were tied up with wire and the robbers escaped with just over $1,100, missing out on close to $40,000 on hand to cover the mine payroll and still locked in the safe. The robbers ran up the narrow path behind the mine office and disappeared into the bush despite yells from Boyle, who had worked loose, for someone to stop them. Two of the three had been recognized as hangers-on in the local area who, until now, had been considered harmless. Mrs. Warren had even called one of them by name as he was tying her up.

Telephone lines to both bank and police station had been cut but, by afternoon, a posse of 20 men led by a constable was scouring the hillside above the mine. The next day they were joined by 30 police, led by Inspector Shirras, and 20 Indians signed on at Shalalth but, by now, the trail was cold. Thursday morning an Austrian, arrested on his way to work at the Wayside mine, was released after disclosing that he had been urged to take part in the robbery but had refused to get involved. As he told it, the trio's plan was to make off with the payroll money, hide out in the hills for a time, and then work as farm hands until the winter lay-off when they would take a boat for Europe.

The robbers could be hiding anywhere in the enclosing sea of mountains but, sooner or later, they were going to have to cross either the PGE rail line or the Fraser River to escape. They almost pulled it off, moving with remarkable speed to cover 40 miles in a little over two days and, if they had been just a few minutes earlier, might well have slipped through Lillooet with no one the wiser. On Thursday evening, Bert Durban, while driving two men out to guard the lower crossing of the Bridge River some five miles north of Lillooet, had startled two men on horseback who swung off the road as if heading for the Indian reserve nearby. The pair were back on the road when he returned and Durban, suspicious, crowded them against the bank, noticing that they were whites, not Indians. With no police left in Lillooet, Durban took over, raising the alarm by telephone and telegraph and organizing a reception committee of vigilantes. Meanwhile, the pair guarding the bridge were approached by a rough-looking man with the toes out of his shoes who told them he was on his way into town to record some claims. Questioned by the guards, he pulled a revolver on them, bounded up the slope into some dense bush and disappeared. The other two apparently slipped through or around Lillooet during the night. A few days later, they were arrested in the Texas Creek area on the west bank of the Fraser River some 12 miles south of town and $1,000 of their haul recovered. The third man simply disappeared.[1]

Tried in early November, the pair pleaded guilty and were sentenced to five years in the penitentiary.[2] With the guilty plea there were no revelations about the identity of the third man, believed to be the brains of the operation, or about their movements in the two days following the robbery.

Pioneer's second bank robbery occurred on May 29, 1942. This time it

The Bank of Toronto branch at Pioneer: robbed in 1935 and again in 1942. Photo by W.A. Hutchings, 1934.

was a one-man job, a risky business when bank tellers still worked from a locked cage with a loaded revolver within easy reach.

It began when a customer, asking to have a word with manager Rush, poked an automatic pistol in his back as they entered the private office. Rush, suspecting something, had stuffed a loaded revolver into his back pocket before showing the man in but had no chance to draw it. Disarmed, he was ordered to call his teller, Knowles, to join them and the latter walked in unaware, leaving his revolver in the locked cage. Covered by the robber, Knowles was forced to tie up Rush using rolls of black tape and was then tied up in the same fashion. Twice, when customers walked into the bank, Knowles was ordered to call out and tell them to come back in a few minutes. The two men were moved to the washroom and gagged with socks drying on an overhead light wire. Returning to get the key to the cage from Knowles' pocket, the robber scooped up about $4,000, walked out the front door of the bank and then, bold as brass, came back along the side of the building and peeked in the window to make sure that all was quiet. Rush worked loose a few minutes later and raised the alarm but, by then, the robber had slipped away unnoticed by mine employees working close by.

Despite a disguise of dyed hair, a Charlie Chaplin moustache and different clothes, Rush recognized the man as a mucker who had quit earlier in the week and left camp planning to find work as a logger. Escape would be difficult; the road past Bralorne could be blocked by a phone call and the high country, still blanketed in deep snow, was probably impassable. The mine manager's police dog, after sniffing a sock discarded by the man, appeared to follow the scent for some distance up the wood stave pipeline along Cadwallader Creek and then, inexplicably, lost it. This suggested that the robber was heading for the high country and the pass beyond, although there was always the possibility of his doubling back along the hillside above Pioneer and Bralorne.

Word of the robbery spread through the valley, but there were no firm leads until three days later when Indians in the Pemberton area, well south of McGillivray Pass, led police to a cabin where the suspect was caught red-handed, parcelling up $1,000 in bills to send to his sister. Exhausted after struggling for more than seven miles through the pass without snowshoes or skis, he was probably glad to have it over. Electing for a speedy trial, he pleaded guilty, claiming to have stolen the money for an operation on his nose, something a doctor had told him was essential to save the sight in one eye. Citing this problem and a crime-free record, he asked for a suspended sentence but, despite this, was probably surprised to receive a two-year sentence instead of the seven years he was bracing himself for.[3]

# *Big Bill Davidson's Minto Mine and Townsite*

Minto, unlike other Bridge River settlements, sprang up because one man, Warren "Big Bill" Davidson, willed it. That will went a long way: one old friend wrote of him, "Big Bill was a tremendous man - I think around six foot three, or maybe more... I don't suppose 'Bill' in actual SIZE was much bigger than many other men, but it was more the way he thought BIG and worked BIG and performed BIG."[1] In anything he attempted Davidson tried to be and often was, the best in the business. In 1934, certain that his Minto mine, under development nearby, was about to become a major producer, he set to work to build the first-class town it deserved.

Big Bill, about 50 at the time, had knocked around the Bridge River since 1912, working as a prospector, miner, carpenter, horse wrangler and woodsman. In 1928 he staked the Minto property and through surface prospecting uncovered a sulphide-rich zone carrying erratic values in gold and silver. The Consolidated Mining and Smelting Company (CM&S), mildly interested but by no means enthusiastic, had taken an option on the property and, as part of the deal, paid Davidson to drive a low-level adit to test the values at depth. Working with one or more helpers and using hand mining methods, Davidson drove about 640 feet of workings before the company dropped the option in 1931.[2] In 1933 Minto Gold Mines Limited, a public company, was formed to develop the property. Davidson, in full control as president, mine superintendent and principal shareholder, plunged ahead with the work.

Big Bill's Minto townsite, financed separately, was developed at a hectic pace during 1934. The site chosen lay on the gravel flats near the mouth of Gun Creek, about a mile from the mine. With the survey complete and the access road opened, trucks loaded with lumber and supplies arrived day and night. Streets were laid out, waterworks installed, and hotel, store and houses started, all in such a hurry that one woman who was there to observe it all was at a loss to recall which came first. By late summer there was even electricity, transmitted over Mission Mountain from the B.C. Electric Company's development on Seton Lake.

On Saturday night, September 29, 1934, a crowd turned out to celebrate the opening of the Minto Hotel. A visitor who had driven down from

Bralorne admired the hotel with its lovely furnishings and large rooms but didn't last long at the party. There was just no room to sit down in the bar and the dance floor was "one pushing-pulling mob of men and women with considerable liquor under their belts although there was no disorder of any kind. As a matter of fact, it was not as sloppy as one of the parties I have witnessed at the Vancouver Country Club… where people are supposed to be more cultured."[3]

At Minto mine a 50-ton-per-day mill was started up November 26, 1934, a mere 77 days after Davidson put a crew to work clearing the site.[4] Late in 1934, the company announced ore reserves of more than 100,000 tons grading over 0.44 ounces of gold per ton,[5] enough to keep the mill operating for five years. Minto differed from Bralorne and Pioneer in that its gold occurred in an irregular fault zone accompanied by masses of sulphide minerals. It was no bonanza; for the moment the operation was close to breaking even. Things might look up if the mill recovery, running in the low 80 percent range, could be boosted and mining methods

The Minto mill and office, 1937, now flooded by Carpenter Lake. Photo by C.E. Cairnes, Geological Survey of Canada, No. 82986.

"Big Bill" Davidson. Photo courtesy of B. Orgnacco.

improved to reduce the amount of waste rock ending up in the ore bins. Resentment against Davidson's one-man show came to a head at the company's annual meeting on September 26, 1935 and, after a stormy session, he was ousted from both board and management.[6]

Certain that the brokers had done him in, Davidson prepared for a proxy battle. His first move was a newspaper advertisement defending his management and asking all Minto shareholders to make certain that shares were properly registered in their own names and not held as street certificates that left voting rights with a broker or his agent.[7] Big Bill's exile ended in mid-January 1936, when Minto's directors invited him to rejoin the board despite objections and subsequent resignation of one director. Then, a few weeks later, he was re-appointed mine superintendent but, on paper at least, his authority was limited to day-to-day operation of the mine itself. He was to have nothing to do with the operation of the mill; Andrew Larson, a fellow director, would plan the development work and generally keep an eye on things.[8]

The Minto townsite, about 1935, now mostly under water. Photo by W.A. Hutchings.

In late February 1936, Larson, a veteran mine operator and consultant, ordered that the low-level or River Tunnel, driven by Davidson for the CM&S company, be extended to test the ore zone currently being mined on three upper levels. In driving the tunnel, Davidson had followed the mineralized zone for 400 feet until it was displaced by a fault and then, turning almost 90 degrees, had crosscut 240 feet to pick up the offset portion lying beneath the upper workings. He had drifted on it for a few short rounds but values were discouragingly low. At this point, CM&S had dropped their option and work was abandoned. On March 7, 1936, a mere 46 feet from the end of Davidson's crosscut, the drift hit spectacular mineralization that assayed 2.4 ounces of gold per ton across the full 5-foot width of the working.[9]

By the end of the day rumours of the new find were spreading through the Bridge River. Not long after, a notice reading "We've hit it at Minto!"[10] was posted on a wall at Bralorne, perhaps by Davidson himself. The find, if it proved genuine, might well amount to something, unlike the fanciful claims that promoters and even their consultants were forever dreaming up in efforts to push a particular stock. Initially, a few of the local people began to buy Minto stock and then, with a few more days of even better news, it turned into a boom unlike anything seen in the valley before. Minto stock which closed at 8¼¢ on March 6 closed at 35½¢ a week later. Many who bought early had already sold and taken their profits, but now, with the drift still advancing in bonanza ore, were wondering about buying in again. Some lucky ones did and around March 23 the Minto boom peaked when a few shares traded at the magic figure of $1.00 on the Toronto exchange. During the flurry, combined trading volumes on the Vancouver and Toronto exchanges totalled about 1.2 million shares a week, or roughly half the total number of shares issued. The banker at Bralorne, ahead $1,600 himself, estimated that there were 60 new cars in the Bridge River, all bought with Minto profits.[11]

In August 1936, when it still seemed possible that the Minto mine could live up to Davidson's expectations, some 300 people made their home in the Minto townsite. In addition to houses, it boasted a hotel, a general store with a 14-suite apartment block above it, a theatre and recreation hall seating 250, a drug store, lumber yard, two restaurants, laundry, garage and filling station, pool room, barbershop, electrical appliance shop, dairy, bakery, tailor and shoe repair shops. *The Vancouver Sun*'s mining editor took it in on a visit to the Bridge River's mines:

> "Big Bill," who visualized all this two years ago says the natives "aint seen nothing yet." He looks for a city of several thousand

The foyer of the Minto Hotel, about 1935. Photo by W.A. Hutchings.

> when Bridge River hits its real stride and to show he believes what he preaches is now putting up half a dozen new dwellings for the two dozen people who clamor for them. And still they come. Yes, Minto City is growing, sure![12]

To celebrate it all, Big Bill organized a Minto Stampede for Labour Day 1936:

> He wasn't content to have another Sports Day, like Pioneer or Gold Bridge. No! Minto Mines must have a Rodeo!! What if he did have to ship in the dozens of horses and all the steers, and import Indians to give it color; erect a grandstand and big outdoor floor for the Indian dances, tent pavilions to feed all the visitors; put up prize money to attract cowboys to ride, etc. etc. Even his Rodeo grounds had to be made, as all of Minto is an old river bed of good-sized boulders. That all had to be cleared of the biggest rocks, and load after load of sand spread to make a footing. The whole thing was fenced and corrals and stands built. All this for a town so new it wasn't even a small dot on any B.C. map![13]

The school at Minto, about 1946. Photo courtesy of Margaret Pasacreta.

Events began with a machine drilling contest held under an arch of rock partly overhanging the road near the Minto mine. Each team drilled for ten minutes, beginning with short steels about two feet in length and changing to longer ones as the hole went deeper. The winning pair from the Minto mine drilled 130¼ inches to win a $50 prize. Next, in the hand drilling contest, it was Algot Erickson, winner of many contests in the Bridge River, but this time with a different and younger partner, Halver Ronning, who drilled 42⅞ inches in 15 minutes to take a prize of $150 and a trophy put up by one of the breweries.

Close to 1,500 people were on hand when the parade formed up at 11:00 in front of the general store. The leaders, all on horseback, were Cal Lee, head of the stampede committee, Curly Evans, Clarence Bryson of Lillooet (the master of ceremonies), together with a squad of provincial police. Rodeo events, including races, bronco, steer and bareback riding, calf roping and wild cow milking, went on over the two days. Cash prizes ranged from $10 up to the $100 won by Jimmy Tegart of Lillooet for the best showing in the two-day bronco riding event.

There were other things to watch, too:

> "Seven fights," chuckled one well-known visitor, "that's what I saw last night - seven of them. Surely that's unusual". Well, yes and no we told him. There have been times when we beheld almost that many ourselves. Never over five, though. Chances are that to make up seven, he counted in the fight card which was part of the program and the one in which "Lay-em-cold" Martley took on, another fair battling bruiser in the first women's boxing match in local history.
>
> ...
>
> At the comparative peace in the suds'-dispensary Monday evening. The largest mix-up of fisticuff on record was in the embryo of this locale when 40 or more of the gentlemen guests got into a friendly little argument over in one corner. It finally spent itself, however.[14]

And:

> The usual quoto of bruises and cuts that ordinarily go with stampedes and the odd fight kept the local medico busy. Perhaps one of the most colorful of the latter was that staged outside the dance hall Tuesday night when one of the stars on the

> afternoon's fight card was knocked cold by a cowpuncher. The cowboy led just one to the chin with his left.[15]

Seven cars were reported off the road on Tuesday morning. No one was hurt, although one car had come to rest upside down about 100 feet below its starting point. "The only one who seemed really excited about it was Vaughn Dubois who, at intrevals [*sic*], was dragged down from his position in the Judges' Box at the Stampede to supervise hauling operations."[16]

Big Bill, seemingly unperturbed over growing uncertainty about his mine's future, held his second Minto Stampede on July 1, 1937, at the same time as the big celebration at the mines. It was less successful:

> A few who, were transported by free truck, went to the Minto stampede, but Dominion Day is usually Bralorne and Pioneer Community Clubs' privilege and a spirit of loyalty and indignation was abroad, that anyone else in the valley attempt a celebration in opposition.[17]

Among those who did go to Minto were the girls from the sporting houses, barred from the mining companies' property aside from their weekly trip. One young man, certain that the stampede was the place to be, hopped the truck to Minto only to find it a one-way trip. The stampede was fun, but he had second thoughts next morning after a night in a haystack and a long walk back to Pioneer with nothing to eat.

Big Bill's Minto mine was fated to have a short, unhappy history. By June 1936, Don Matheson of Bralorne had taken over as general manager of the operation, apparently with an understanding that he could return to Bralorne if things didn't work out. A winze* or inclined shaft was sunk for 143 feet down the dip* of the mineralized zone and a new 5 Level driven. The rich, heavy sulphide ore, drifted for close to 300 feet on 4 Level, petered out 35 to 40 feet down the winze and, by the end of September, some 350 feet of drifting on 5 Level found only some sections of sulphide concentrations but no continuous showing of high-grade as on 4 Level.[18]

Despite poor results on 5 Level, Matheson had big plans for Minto: too big, as it would soon turn out. Late in 1936, work began on a three-compartment shaft to test the mineralization down to the 7 Level, some 400 feet below the rich ore on 4 Level. While this was being sunk the mill continued to treat close to 100 tons of ore per day. When Minto Gold Mines held their annual meeting on October 29, 1937, Matheson was already back at Bralorne, and Minto's president, William Warner, acknowledged

School children, some of them Japanese-Canadians, outside the Minto school, probably 1946. Photo courtesy of Margaret Pasacreta.

that the company, having depleted their finances in the shaft sinking venture, was now working from hand to mouth. [19] The shutdown came on December 19, 1937, at which time there was little ore in sight and the stock was going for a mere 3¢ a share. In total, including later brief attempts at revival, some 87,100 tons of ore were milled and 17,557 ounces of gold and 50,583 ounces of silver recovered.[20]

The settlement of Minto hung on after the mine closed. It became more of a ghost town each year, with the except for a few months in the spring of 1940 when Big Bill Davidson and Curly Evans leased the mine in an unsuccessful attempt to put it back into production. Suddenly, in May 1942, Minto was full again, but this time it had nothing to do with mining. Its new residents were Japanese-Canadians forcibly removed from the coast during the wartime emergency. About 25 families were involved and, despite some initial opposition, were soon accepted into the community. Many of the men found employment in logging. Until their departure following World War II Minto blossomed with spectacular gardens developed on the barren gravel flat.

# *Summer Celebrations and Hockey Rivalries*

On three summer holidays, Victoria Day, Dominion Day and Labour Day, the mines were shut down and happy crowds turned out to make the most of the occasion.

## *Victoria Day, May 24*

This holiday came into its own in 1936 when miners and the town of Gold Bridge got together to celebrate the first anniversary of the return to work following the 1935 strike. All the mines in the valley took the day off and the men flocked into town. The day began with a visit to the cemetery to lay wreaths on the graves of two miners killed at the Pilot mine and another two killed in a slide at Ogden. Brief remarks by Lars Larsen, leader of the miners, were followed by a half-minute of silence, probably the only quiet period of the day.

Morning events included the single-jacking contest, won by Federal Mine's Algot Erickson, a man in his late 50s, who drilled 14⅛ inches in the allowed 10 minutes, close to the world record of 15 inches set in Australia. Bralorne's baseball team played Minto's in the morning and Pioneer's in the afternoon, beating both. The evening fight card was one of the highlights:

> The valley pugilists laid the fans in the aisles, the ditches, running boards and eaves-troughs as they flim-flammed each other to the great glee of every gentle feminine heart in the spectators benches. Scores of on-lookers hung out of the Royal Hotel windows to the peril of the crowds below; many chose the roof of the News Office, the Model Bakery and Traveller's Cafe. "Give him hell - kill him . . ." yelled the sweet young things while their masculine escorts indulged in a little of the "fisticuffs" to settle their own little grievances outside the ring.[1]

There were several thousand people in Gold Bridge that day and, at the Royal Hotel, Johnny Branca began selling draft beer at 10:30 in the morn-

Tug-of-war at Sports Day in Gold Bridge, May 24, 1935. Photo courtesy of B. Orgnacco.

ing and kept at it until 11:00 at night without as much as a break to eat. It was a thirsty crowd; Branca's take at two glasses for a quarter was a record $998 and Branca himself would have thrown in the extra $2 to make the thousand had he known the exact figure at the time. By and large it had been a happy day and, despite high spirits, the crowd had been an orderly one.

The 24th of May holiday continued as Gold Bridge's day for the next few years. But things were quieter, the big crowds and the excitement gone as the mining camp settled down to two profitable mines and only limited activity on a few other properties and prospects.

In 1939, Gold Bridge opened a new park developed on land donated by Ernie Shepherd's B.R.X. Mines. George Murray, the local MLA on hand to cut the ribbon, made a speech predicting brighter days ahead for the valley. The customary races and ball games were held plus a new event, a banana-eating contest. In the midst of it all, it took three hours of trying for Ernie Howard of Minto to make it to the top of the greased pole and grab the $5 bill fastened there, others giving up and settling for the $1 and $2 bills lower down. Howard needed the money for his miner's licence, the reporter noting that "his earnings have been very slim so far this spring."[2]

## Dominion Day, July 1

July 1, called Dominion Day in those days, was the big one and, although celebrated since 1928, it really came into its own after the mines built ball parks in the mid-1930s and, soon after, switched to a six-day work week. Beginning in 1934, Bralorne hosted the celebration in the even-numbered years, Pioneer in the odd. There was rivalry between the two mines, but on this day it was more in the spirit of fun than the blood-letting of the hockey season. Teams and individuals from the lower valley took part as well and up 1,500 people turned out. Programs included the usual foot and novelty races for men, women and children; one or more baseball games; a tug of war and mining contests. On the practical side there were fire drills with generous cash prizes put up by insurance companies. Both mines had their own bands, Pioneer taking great pride in their 33 piece brass one.

Hand-drilling contests were the highlight of the mining events in early years, something practical since the method still in use at some of the smaller properties. In the contest a team of two men (double-jacking) or, less frequently, one man working alone (single-jacking), would compete to drill the deepest hole in a slab of rock set in place for the competition. In double-jacking, one man would hit the rock drill with an eight-pound sledgehammer while his partner, holding the drill, would keep turning it to keep the hole clear of rock dust. When the man hammering tired, the two would switch jobs in a precision movement. In the 1934 contest, Frank Elsener with Charlie Lougheed won $100 each, almost a month's wages, by drilling 28 inches in 12 minutes. With the exception of one team with a mere 18¾ inches, four other teams were within 4 inches of the winners, one a scant 3⁄16 of an inch behind.[3]

The mucking contest was another popular event. Two men shovelled ore into a one-ton car, pushed it up a short length of track, threw a switch over and back, returned the car to the starting point, and dumped it. The time taken, usually three minutes or less, depended on the layout used each year and often only a few seconds separated the top teams. Sometimes teams handicapped by too many drinks earlier in the day would lose out, throwing shovelfuls of muck right over the ore car in sheer exuberance. The ladies competed in a separate event, but their times seem to have gone unreported other than to note that they were "good."

In the evening, a dance sponsored by one or both of the community clubs wrapped up festivities. The dances were dry, with no liquor allowed in the hall under the laws of the time. Drinking went on beforehand and, during the evening, some of the men and the occasional woman would slip out to the parked cars to gulp raw liquor straight from a bottle. It had to be done with care; the police, well aware of what went on, were ready to arrest anyone too obvious about it. In 1938, one of the bigger years, the two community clubs went all-out for the dance, with two local bands, Pioneer's Valley Vagabonds and Bralorne's Benny Stone and His Royal Commanders, playing in alternate one-hour shifts far into the night.

## *Labour Day*

This holiday, held on the first Monday in September, traditionally wrapped up the summer's activities. In 1934, Gold Bridge hosted a miners' exhibition. The main event was a hand-drilling competition with the target a huge granite boulder left behind when one of the lots in town was excavated. Close to a thousand spectators looked on as Algot Erickson and Mickey McLennan drilled $37\frac{3}{16}$ inches in 15 minutes to take the $150 prize. They were an unlikely team, Erickson, a towering, powerfully-built man and McLennan, small and wiry, but, despite that, their teamwork was

Erickson and McLellan in the hand drilling contest at Gold Bridge, September 3, 1934. Photo courtesy of R. DuBois.

Coronation of a Snow Queen at Bralorne Community Hall, mid 1940s. Photo courtesy of the Baxter family.

near-perfect. Swinging an eight-pound sledgehammer the pair used and discarded some 14 steel bits of varying lengths.[4]

On Labour Day 1935 the Second Annual Sports Day was held in Gold Bridge. This time there were the usual foot races as well as three-legged, sack, potato and wheelbarrow races. In the hand drilling it was Erickson again, this time with another partner, B. Tramm, drilling 38¾ inches in 15 minutes to beat out a Bradian team by a mere half inch. In baseball, Pioneer outplayed Bralorne in two straight games to take the Bridge River championship with a total score of 22–4.

In 1936, the Minto Stampede took over as the valley's Labour Day event and, by 1938, the celebration was down to a sports day for the children put on by the miners' union at Bralorne. Even this was jinxed, with the Pioneer children, quarantined for chicken pox, stuck in their own camp and unable to take part.

## Hockey

Sports, competitive and otherwise, were an important part of life in the Bridge River. Bralorne and Pioneer, joined at times by Minto or Gold

Bridge, competed in baseball, basketball and, most of all, in hockey. Determined to win, both mines used offers of good jobs to recruit some of their better players. Word got around and other would-be athletes came to the valley in hopes of trying out and winning both a place on a team and a job. Much of the time the Bralorne and Pioneer communities went their separate ways but for an important game a mob of fans from the visiting team would make the 2½-mile walk to support their boys.

At times the hockey rivalry came close to getting out of control. Games were played outdoors and keeping warm could be a problem for the spectators. One dodge was a thermos full of hot rum, tucked under one's overcoat and tapped by a thin rubber tube such as a catheter "borrowed" from the hospital supply. Others kept their circulation moving by reaching out and trying to hit "bad guys" on the opposing team when they skated close to the boards. At one game, some overzealous women pushed too hard and 30 feet of the boards collapsed onto the ice. Occasionally, a game would get completely out of control as the fans joined the players for a wide open free-for-all on the ice. The recollection of one spectator, a teenager at the time, sums it up: "Mother used to try and spit on the [opposition] players."

Both mines imported hockey players and gave them jobs, the latter not too onerous and certainly nothing that would get in the way of turning out for hockey practices. At one time Bralorne had a crew of players retimbering portions of the old King Mine with the hockey club president as foreman. It kept them clear of the active part of the mine and no one seemed too concerned over reports of sleeping on the job. Sometimes, before an important game, a hockey player who did have a more demanding job would be helped out by off-duty fans, anxious to see him turn out in top form.

At Pioneer, Ed Emmons, the mine manager, was the hockey nut while at Bralorne his counterpart was Ted Chenoweth, the mine superintendent. Neither were the top men and their bosses, James and Bosustow, respectively, did their best to keep the rivalry within acceptable limits. Even so, almost anything went and at one point Bosustow threatened to stop the games altogether unless there was an end to the bloodletting.[5]

The first hockey league was organized in December 1935 shortly after both mines completed their hockey rinks. For the moment, there were teams from Pioneer and Bralorne with provision for a team from the lower valley to join in at some future date. Charlie Cunningham, the well-liked insurance man from Gold Bridge, was league president, and the local bank managers were persuaded to act as treasurer and secretary. It was just as well they did not have direct connections with the mines, as they were soon embroiled in dealing with disputes. There were differences over who was eligible to play under the B.C. Amateur Hockey Association rules and over the nature of the ice at the two rinks, the latter almost unavoidable with Bralorne's rink on a sunny bench and Pioneer's deep in the valley of Cadwallader Creek and cut off from the sun until mid-February. And Bralorne backers maintained that Pioneer's practice of flooding their ice with hot water just before a game to produce a smooth skating surface was, in reality, just a dirty trick that gave the home team an unfair advantage.

Both sides were curious about new players imported by the opposition. Here Bralorne had the advantage. A new player headed for Pioneer had to pass through the Bralorne camp and, if he travelled on one of Curly Evans' stages, word was almost certain to reach Bralorne where Curly had close ties. There is even a claim, perhaps imaginary, that one new player headed for Pioneer was met at Shalalth by the Bralorne group, wined and dined en route, and signed on at Bralorne before he realized where he was.

Spring ski outing on Sunshine Mountain south of Pioneer. Photo courtesy of the Baxter family.

All hockey players had to meet B.C. Amateur Hockey Association rules with regard to amateur status, residence requirements, etc. There were games to be played here as well as on the ice. The opposition was always curious about a new player, certain that somewhere in his background there must be grounds for a protest to association headquarters in Trail. The closer the competition, the less the likelihood of any give-and-take. Both sides were out to win and if an opponent could be sidelined either by rules or by injuries, so be it.

The first exhibition game was played at Pioneer on New Year's Eve 1935, Pioneer taking it 6 to 1. It was followed by a dance that, for some at least, lasted well into the next day. Bralorne, outclassed, moved to strengthen its team and, before long, close to a dozen new players appeared, all looking for a place on one of the teams. Despite changes, Bralorne, off to a bad start, lost the first two games and tied a third. Further strengthened and reorganized, Bralorne was out to win a game on their home ice on Sunday, January 26, 1936:

> Quite a brilliant game, but unfortunately in the third period, the game got out of control. The players began to mix it up, started by Gough's [Pioneer] rough work. Pretty soon the spectators climbed on the ice and a regular melee was in progress. Luckily no sticks were used, but there were a number of punishing scraps. Pug Hayden [Bralorne] distinguished himself by some pretty hefty work. The game was finally resumed, but fighting took place on the side lines and even when the game finished as a 1–1 tie, more scrapping took place around the dressing rooms. Hayden punished Red Heckler [Pioneer] much to everybodys enjoyment. Too bad as it is apt to leave some nasty feelings around the country. Hear that Bud Powell's [Pioneer] jaw was broken in yesterday's imbroglio. . . . he has to go down to Vancouver for treatment.[6]

After six games, it was Pioneer with three wins and two ties, going into the "supposed playdowns"[7] to represent the Bridge River in competition for the B.C. Intermediate Hockey League's Coy Cup. Unexpectedly, Bralorne took the first game 3–1 and, with that, Pioneer protested that some of the Bralorne players did not meet the residence rules. After frantic telegrams to association headquarters in Trail, Bralorne produced "some kind of OK" of questionable value for Craigen, their goalie.

There were still problems. By one count Bralorne had only eight eligible players, and a hectic debate at an executive meeting delayed the start of the second game for more than an hour, leaving the fans to stand around in below-zero weather. Finally, Pioneer agreed to play under protest, winning the game 5–2. An incident in the game added to the bad feeling in both camps: Gough of Pioneer deliberately skated down the ice, hit a Bralorne player over the head with his stick and then tossed the stick out over the crowd as he skated off the ice.

Pioneer, declared the winner, was awarded the Murray Cup for hockey competition in the Bridge River valley. The cup had been presented by George Murray, local member of the provincial legislature, and a column in Murray's *Bridge River-Lillooet News* sniffed: "The Murray Cup... has been won and lost for first time, not on the ice where it was intended to be won, but in a committee room."[8]

For the Pioneer team, the next stop was Merritt and the playoffs for the Coy Cup. Unable to take the entire team, Pioneer was badly outclassed, losing two quick games to the Merritt team in late February 1936. In all, it had been quite a season and Bill Hutchings, the league treasurer, noted in his diary: "Next year, I will try and keep my head out of the noose of athletics."[9]

In the 1936–37 season, a third team, representing the lower valley from Fish Lake (Brexton) to Minto, was added to the league. But the main contest was still between Pioneer and Bralorne when the teams faced off again early in 1937. This season it was unequal; the Pioneer team, larded with four players from the Merritt team that had beaten them in last season's battle for the Coy Cup, won the Bridge River championship without losing a game. Next, in the playoffs for the Coy Cup they defeated Wells at Pioneer, Merritt and Princeton in games played at Merritt, and finally Salmon Arm on that team's home ice to take the cup. On their return to Pioneer, the company put on a banquet for the players followed by brief speeches and the presentation of 15-jewel gold watches to the players, manager and coach. Then, master of ceremonies Rev. S.H. Smith tactfully left for a real or imaginary engagement at Bralorne while the victory celebration continued in the Number Three bunkhouse.[10]

In the 1937–38 season, Bralorne got serious about hockey. Most of their players were new arrivals and the team had a new name, the Bralorne Golddiggers, with accompanying flashy gold and black uniforms. Their coach, "Speed" Wilds, had been captain of the team for the past two seasons. The Bralorne fans came up with their own song:

In the good old winter time
In the good old winter time
Skating down the Bralorne ice
With the forward line
We pass the puck, we shoot it in
And that's a very good sign
That we're going to win the Stanley Cup
If you'll only give us time[11]

In their first game with the Pioneer Highgraders, the teams fought to a 6-all tie, both scoring a goal in overtime. But next day, Bralorne took the game 2–1 in a stormy session that, at one point, left only three men on the ice for each team. From then on the Bralorne Golddiggers dominated, winning or tying all but one of the games with Pioneer. By the end of the season the league was back to two teams, the lower valley team gone, a victim of financial problems.

After winning in the Bridge River, the Golddiggers moved on to represent the area in the Coy Cup competition. There was no stopping them; Merritt was beaten in two quick games and Vernon two games to one.

The Bralorne Golddiggers, winners of the Coy Cup, March 1939. Photo courtesy of Irene Howard.

Hockey fever gripped the Bralorne camp and, after the first Vernon game, a contingent of excited fans set out to watch the remaining games. Others, unable to get away, huddled around the radios in camp twisting dials in hopes of getting snatches of the game broadcast on the Kelowna station. Total score for the Vernon games was 20–7 and the Vernon News described what hit the home team:

> The [Bralorne] club's... roster reads like a National Hockey league team's, members having been drawn from Vancouver to Halifax. The three masterminders directing from the bench - Victor O'Brien, manager; Walter Hayden, trainer; Norman Wilds, coach - hail from Halifax, Vancouver and Coleman, Alberta, respectively.
>
> The defence includes two goalies, Gordon Craigen of Lacombe, Alberta, and Jackie Hutton, once of Kimberley Dynamiters. Others are Art O'Keefe, Sudbury, Ontario, Scotty Skuratoff, Vancouver, and Art Moffatt, North Battleford, Saskatchewan.
>
> The first string forwards, "Boyde" Clarke, Vern Kunsman and Mike Purcello have previously played in Calgary, Portland and Rossland. Lethbridge is the home town of Tommy McDowall, Calgary of "Jiggs" Gooderich, and Sudbury of "Bus" Woods.
>
> Bralorne lost but one game all season, to Pioneer by a single goal.... A few of the players have seen and competed in a lot stronger competitive hockey in days past and everyone on the club has at least seen action as a senior.[12]

Following the Coy Cup win, the Bralorne Golddiggers went on to defeat the Alberta and Saskatchewan champions, the Canmore Briquetteers, 9–8 in a two-game, total score series to take the Western Canada Intermediate hockey title.

In the 1938–39 season, Bralorne and Pioneer were at it again, both determined to win. Pioneer had a new coach, "Dutch" Gainor, a former defenceman for Boston in the NHL, who started practices before the first of December. One observer commented, "If the boys could only master a small fraction of the knowledge and advice that Dutch is handing out they would be a team second to none in the province."[13] There were risks too; Cassidy, the Pioneer goalie, hit by a flying puck was out with a broken jaw after the first practice.

The Bralorne Golddiggers, stronger than anticipated, walked off with the Murray Cup for the Bridge River with nine wins, three losses and a single tie. Bralorne scored 64 goals out of a total of 469 shots on the Pioneer goal. Pioneer had 45 goals in 300 shots. Bralorne led in penalties, too, with a total of 244 minutes off the ice compared to Pioneer's 214 minutes.

Pioneer, far from finished, had saved a surprise for the three-game series to decide the contender for the Coy Cup:

> Beyond the wildest calculations of the fans and the fondest expectations of their coach, "Dutch" Gainor, the Pioneer Highgraders out-played, out-skated, out-teamed and out scored, in not only one but two straight games on respective rink last Saturday and Sunday, when Bralorne Golddiggers, champion holders of the Murray Cup went down in defeat....
>
> After the shouting of the victors died down and the post-mortems of the vanquished cooled off, there could be no doubt in the minds of the fans, but that "Dutch" had something up his sleeve after all. Much credit was given to Art Gagne [the referee brought in from Kamloops] but he could not put that "surprise" stuff in Pioneer's playing. It took more than any clever coach such as Gainor, to bring out a score of 4 to 1 in the first game and 8 to 2 in the second, even plus the alibis about ice conditions, rink location and what have you.
>
> And did the people like it! Nearly one thousand turned out, paid fifty cents apiece and stood in the biting cold; enjoyed it; yelled for more, and will part with a dollar rather than miss the third game on next Saturday night. It was the kind of hockey the fans often read about. It was fast and clean and ... believe it or not, in this gold camp, the first game without a single penalty.[14]

The third game was played at Bralorne the following Saturday "despite pelting snow that lost the puck in a heap before it crossed the blue line. The players could have passed for golfers as they copped the snow back and forth. The spectators, about 500 got a good soaking. However, most of them stayed and took it."[15]

Pioneer lost the game and, even worse, learned that their first two wins had been disallowed on a ruling from the hockey association headquarters in Trail for reasons still unknown. Indignant, the Pioneer team refused to go out on the ice for the game scheduled for the following afternoon, leaving the executive of the Bridge River Valley League to cope with the situation. That group, together with referee Art Gagne, met all afternoon and far into the night, finally adjourning without a solution.

On Wednesday, the league came up with their official decision. From now on it was to be their show; all outside interference and rulings would be disregarded. Of the earlier games, the first had been ruled out on the grounds that Moffatt, a Bralorne defenceman, had been suspended pending a ruling on his case; the second awarded to Pioneer; the third ruled out on the grounds that Trites, a Pioneer forward but formerly with Bralorne, had been suspended without their consent and the fourth, scheduled but not played, awarded to Bralorne by default. That left the two teams tied 1–1 with the two games scheduled for the coming weekend and another game, if necessary, at a date to be set by the league executive.

Bralorne "Cubs" Hockey Team, 1939–40.
Standing: L. MacKenzie, u; J. Beck, trainer, A. Lougheed, manager; E. Menhennick, r.w.; P. Ashmore, def.; E. Smith, def.; J. Lougheed, c.; W. O'Keefe, coach; M. Sherlock, l.w.
Seated: C. Henderson, l.w.; G. Ciaco, def.; J. Markin, c.; R. Movold, goal; K. Wilson, l.w.; A. Frydenlund, def.; B. Movold, r.w. Photo courtesy of M. Jukich.

By the weekend it changed again, this time to a two-game, total score series to permit referee Art Gagne to leave for Kamloops on Sunday evening. It was anticlimactic: the Bralorne team, recovering from the shock of the earlier Pioneer offensive, "never played better hockey and their fast skating, smart stick handling, superb back-checking and fine team work showed up to advantage against the Pioneer team who lack the organization they displayed" in the earlier games.[16] Bralorne, the winners 13–6, were once again on the way to the Coy Cup.

On the road, the Bralorne Golddiggers bested New Westminster and then Merritt in two-game, total point series to put them in the finals at Vernon playing the Trail team. The first game ended in a 2–2 tie, despite three overtime periods; Bralorne squeaked the second 2–1 and, in top form, took the third 6–3 to retain the Coy Cup. That was it; riddled by injuries in the playoffs, Bralorne lost to Edmonton in the semi-finals of the Western Canada Intermediate playoffs.

There was no 1939–40 hockey season. The serious rivalry was over. Pioneer was shut down by a strike and, the following season, an attempt to start up again fizzled out despite the comment in the *Bridge River-Lillooet News* of November 22, 1940:

> Ed Emmons and Ted Chenoweth are getting all set to begin the old Hockey argument. Ed and Ted don't go at it face to face, but they do a lot of primin' from the respective bleechers [*sic*]. There's no arguin' about how successful they used to be.

With wartime shortages of manpower, the luxury of a hockey team was out of the question and over the next few years, interest gradually shifted to the gentler sport of curling. In fact, about the time the comment was written, Pioneer was completing a curling rink with two sheets of natural ice.

# *The Later Years*

*1942–1971*

At a party held for Dr. King, on a visit to Bralorne before going overseas, March 1944. From left to right: McDermid, Echlin (in rear), Cameron, Ledger, C.P. Ashmore, S.S. Railton, and Dr. King. Photo courtesy of Mrs. D.M. King.

# *Wartime Changes, 1942–1945*

By late 1942 the Bridge River was no longer the place to be. The excitement was over and the war that once seemed so far away was now the main concern. Many of the younger men joined the armed forces and others, too old or medically unfit, drifted away to work in the shipyards and other wartime industries springing up in the Vancouver area. Some family men, still bitter over the Pioneer strike, had wanted to leave for years, but for others a petty matter like a run-in with a shift boss was all it took to start them down the road.

## *The Kings Leave Bralorne*

For many, the war and what it meant hit home in late June 1943, when Dr. King, after 11 years at Bralorne, left to serve in the army medical corps. A monster surprise party was organized and $1,000 in individual contributions matched by the company used to buy gifts, a set of instruments for him and a sterling silver tea service for his wife Torchy. Escorted into the hall by the Mathesons and the Ashmores, the Kings were overwhelmed by the crowd and the decorations, in particular by a large stork carrying a baby and the sign, "Brought three hundred babies into camp and never lost a father."[1] Bert Ashby, the amalgamator and a holdover from Lorne Gold days, was master of ceremonies, and during the speeches there wasn't a dry eye in the hall. Jackie Muir, the first baby born in the Bralorne camp, presented the tea service. As Dr. King left the stage, some of the men lifted him to their shoulders and paraded him around the hall.

The Kings, accompanied by Mrs. Ashmore, slipped out of Bralorne soon after. The news travelled fast, and, all the way to Shalalth, knots of people waited beside the road anxious to say good-bye to the couple and to wish them well.[2] One of the strongest links binding mining camps and settlements together had been broken and, although things would appear to go on much as before, the Bridge River would never be quite the same again.

## Gold Mining

During the war, gold mining had a low priority. In the United States, many gold mines were shut down in December 1942 by government order and, in Canada, "essential" industries had first call on manpower, equipment and supplies. Master mechanics and shop crews at both mines showed great ingenuity in keeping old, worn-out equipment running a little longer and in making do with materials at hand. One by one, gold properties that both Bralorne and Pioneer had attempted to develop outside the Bridge River area were shut down as exploration efforts concentrated on strategic metals, particularly mercury and tungsten.

During 1942, Bralorne's underground crew dropped from an average of 294 men per day in the month of January to 137 in December, with the average for the year 211, down from 348 in 1941. Pioneer, hit harder, saw their underground crew drop to about 60 men by the end of 1942, roughly a third the number needed for an efficient operation, and daily production dropped proportionately from 300 to 100 tons per day.[3] Pioneer had fewer opportunities than Bralorne to cut corners; most of their crew had to be put on maintenance work and, even so, an ever-increasing amount of retimbering would have to be done before ore, already blasted and lying broken in the stopes, could be recovered and milled.

There was a job for anyone willing to work as the companies struggled to maintain production with skeleton crews that, by and large, lacked mining experience. Beginning late in 1942, farmers from Saskatchewan were lured to the mines with the promise of a few months work in what was normally their slack season. The companies, well pleased with the results, continued the practice throughout the war years. Even a day or two could make a difference, and in one instance a Fuller Brush salesman who had done well around the Bralorne camp was persuaded to spend a few days on the end of a muck stick while waiting for his order to be delivered.[4] Women took over some of the mine jobs, but the companies never defied tradition and employed them underground. Some miners, Cousin Jacks among them, considered women underground an evil omen, certain to bring down disaster and, if the issue had been forced, might well have walked out or quit.

There were bright notes at both mines. Bralorne prided itself on maintaining or even increasing its reserves despite a drastic reduction in development work. Pioneer, unable to maintain either gold production or reserves, had a future potential its new 27 Vein which, although discovered earlier, first showed promise when explored on 20 Level in 1941–42. Both mines struggled through the remaining war years, Bralorne's bullion production falling to just over half pre-war levels, while at Pioneer production fell steadily to a low of 4,317 ounces of gold in the year ending March 31, 1946, not quite 5 percent of their best output some 11 years earlier.[5]

## *Paddy the Bootlegger*

During the war years, beer and liquor joined a lengthy list of items subject to government regulation and rationing. Despite shrinking mine crews, the Mines Hotel could only get enough beer to keep The Big Stope open about half the time, and stocks at the government liquor store were often low or non-existent. Private enterprise in the form of bootleggers and moonshiners expanded operations to meet the demand.

J.P. Ryan, better known as "Paddy the Bootlegger," was an old hand at the game. Almost everyone in the Bridge River knew what he was up to and where to find him in an hour of need. Originally a desert rat from California, his police record went back to 1934 when, on two occasions, he had been raided, had his moonshining outfit destroyed and been fined $100. Paddy's favourite spot to pitch potential customers was in the "Mens" at the Big Stope. His product came in three grades, Nos. 1 to 3, all distilled from the same batch and priced according to "quality."

He seldom touched the stuff himself and when he did go on a bender it was done in style; stocking up with rum, whiskey and Logana (a cheap loganberry wine) at the government liquor store and taking a room at one of the hotels. The good stuff was for Paddy and the Logana for customers who dropped in on him. Regardless of the weather outside, Paddy kept the room window open and slept on top of the bed, fully dressed right down to his rubber boots.

One hot summer day when John Branca was stacking beer cases at his Mines Hotel, two well dressed men who had checked in a few hours earlier came up and began questioning him about bootleggers and stills in the area. Obviously police, they wanted to know if Branca knew Paddy Ryan. Branca allowed that he did but suggested if Paddy and perhaps others were up to something it was on a very small scale and not hurting anyone.

At suppertime, Branca's wife asked him if he had seen the two lovely police dogs in the pair's car. Branca realized these were the dogs described in a recent newspaper article as trained to sniff out moonshine and stills. Branca, on good terms with Paddy, hung around in front of the hotel for several hours, hoping someone trustworthy would come by to take word to Paddy to bury his moonshine well away from his Gold Bridge shack.

Unfortunately, he chose the wrong man; the message got through but no action was taken when the pair went on a bender that lasted until 4:00 a.m.

Two hours later one of the policemen pounded on Paddy's door, and Paddy, sleepy and hungover, stuck his head out and gave him an earful. That did it. The other policeman and the dogs were called over and the moonshine in Paddy's yard turned up in short order. Paddy, charged with possession, landed in the Bralorne jail and phoned Branca for help with the $200 bail. Branca, fed up with the turn of events, called Paddy a knucklehead who deserved everything he got and, for good measure, added he hoped the police would throw away the keys.

Everything had gone wrong. The police had only wanted to get a look at Paddy, not run him in, and when Branca relented and bailed him out a contrite Paddy could only reflect, "Dogs are my best friends and they had to go and betray me to the cops!"[6]

# *Gold Mining: A Troubled Industry, 1945–1958*

By late 1945 ex-servicemen were returning and Bralorne and Pioneer began rebuilding their operations. Experienced miners were still in short supply and the mines, once preferred places to work, now had to compete with demands from an expanding base metal mining industry. Any ideas about a return to their pre-war style of management ended when the International Union of Mine, Mill and Smelter Workers called a province-wide strike on July 3, 1946.

Following their 1940 defeat at Pioneer, the Mine-Mill union had regrouped. About 1944, it won government recognition as the bargaining agent at both mines although, as yet, it had been unable to negotiate a collective agreement with either company. Early in April 1946, the union's Bralorne local presented management with Mine-Mill's Eight Point Program calling for a 40-hour week with a 29¢ an hour increase in pay, pay for statutory holidays and vacations, plus other benefits including the check-off for union dues. A counter offer by Don Matheson on behalf of the Bralorne was rejected. The local held out for their eight points, unwilling to negotiate for anything less.[1] Mining continued for another month on a much reduced scale as miners, anticipating a long strike, left to look for work elsewhere.

When the strike was called on July 3, both Bralorne and Pioneer shut down immediately, and this time no attempt was made to break it. Staff members continued to be paid, and the union, for its part, supplied the maintenance men needed to keep the mines pumped out and the electrical and mechanical equipment in good working order. Later, in Bralorne's 1946 Annual Report, Matheson wrote of their "full co-operation" in this regard. There was more of the same behind the scenes; unknown to Bralorne's directors, their company store had carried some of the striking miners.

The two major copper producing mines in the province settled in mid-October 1946, but the strike at the gold mines dragged on until late November. By then, both sides had had enough and new recommendations by Chief Justice Gordon Sloan of the B.C. Supreme Court gave them a chance to settle without losing face. For the union, an additional 20¢ a shift brought the total increase to 84¢ for each eight-hour shift. In return, the union accepted conditions put forward by the mining companies that carried both obligations and, hidden in some tricky wording, the companies' acceptance of the distasteful "check off."[2] Bralorne reopened on November 25, 1946, and Pioneer two days later but the latter, losing much time in rebuilding the crew and catching up on timbering, did not resume milling until January 28, 1947.

By now gold mining was a troubled industry, selling its product for a fixed price while operating costs rose steadily. The $35 US price for a ounce of gold, a godsend in 1934, was turning into a disaster. That was bad enough but there was government interference too. On July 5, 1946, the Canadian dollar, pegged in wartime at about 91¢, was arbitrarily boosted to par, in effect reducing the price of gold from $38.50 to $35 Canadian.

In the years ahead, mining industry spokesmen would flip-flop between predicting that governments were about to come to their senses by increasing the price of gold and by begging for special assistance for gold mines. The latter arrived in 1948 with the Canadian government's Emergency Gold Mining Assistance Act which paid a portion of production costs in excess of the fixed price of $35 US an ounce. Inflation was the real culprit, though, and from time to time governments would talk of fighting it, but it was mostly just that: all words and no action.

Exploration at both mines was encouraging; had it been otherwise they could well have shut down until governments raised the price of gold. At Pioneer, ore reserves as of March 31, 1949, were estimated at 426,000 tons grading 0.50 ounces of gold per ton, up from 118,524 tons of 0.467 ounces a year earlier, as reserves from the new 27 Vein were included for the first time.[3] Bralorne had done well between 1943 and 1948, maintaining ore reserves on the order of a million tons carrying slightly more than half an ounce of gold per ton. Almost all of the Bralorne reserves were on the former Bradian ground where, in 1940, the Empire and Crown Shafts had been sunk to 20 Level in preparation to mine the 900-foot block of ground below 14 Level, the former bottom of the mine. A quarter of the gold produced from 1943 to 1948 came from the Coronation (77) Vein, unsuspected until discovered on 20 Level in 1940 and later found to be ore-bearing to just above 14 Level.

Both mines struggled to keep mining costs down. With wages increasing steadily, the best way to do it to use machinery to replace muscle power and speed up the work. One major improvement in the early 1950s was the introduction of the lightweight jackleg drill with throwaway bits. The new drills, light enough for one man to handle alone, were braced with a single, adjustable leg in place of the heavy steel bar that had to be wedged in position prior to clamping the old-style machines on it. Tungsten-carbide–tipped throwaway bits eliminated the need for a steel sharpening shop on surface and cut down the amount of drill steel handled up and down the shaft. There were other changes too. In the stopes, scrapers operated by compressed air were used to move broken ore to the chutes and, when driving new crosscuts and drifts, mechanical mucking machines,

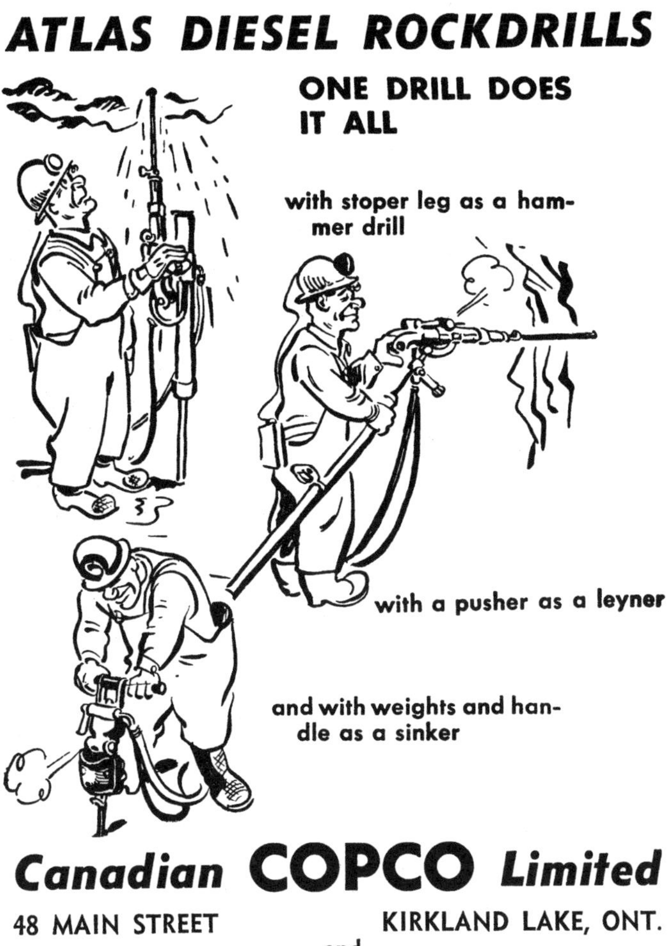


New equipment such as these drills helped improve efficiency in the mines in the 1950s. This advertisement appeared in the *Western Miner* in 1950.

also powered by compressed air, were used to load the broken rock into ore cars. On many levels, small, battery-powered locomotives replaced the slow work of hand tramming. Another innovation was the use of rock bolts, wedged into drilled holes and tightened, eliminating the need for some timbering. At Pioneer a sand-preparation plant was built in 1954 so that sand extracted from the mill tailings could be pumped back into the mine for use in backfilling* stopes.

Financially, neither company was doing particularly well despite government aid and attempts to cut costs through mechanization. Pioneer was hardest hit. After paying dividends of 80¢ a share in the year ending March 31, 1943, dividends dropped to 13¢ in the next year, then nothing until a payment of 10¢ a share a full decade later. Bralorne did better, paying $1.20 a share for the calendar years 1939 though 1944, dropping to 40¢ in 1946 and suspending dividends in mid-year. Dividends resumed in 1949 with an interim payment of 10¢ on January 15 and, with the exception of payments missed for a three-year period beginning in late 1952 and for two quarters in 1957, maintained at 10¢ a quarter until the end of 1965.

## *High-Grading Again*

Although less of a problem in later years, high-grading could still happen. The most bizarre case in the Bridge River took place at Bralorne in 1949. It began Easter Sunday morning when Don Matheson, the general manager, was called aside while watching the children's Easter egg hunt on his lawn and told that seven segments of gold sponge, worth about $14,000, had been stolen from the mill.[4] The sponge, a porous mass of gold and alloyed silver remaining after mercury is retorted off an amalgam, had disappeared from a strongbox kept inside the locked cage where the amalgam was handled.

Security was sloppy. Everything that went on in the cage could be observed through a glass partition; one of the two doors to the cage could easily be sprung with a penknife or a thin strip of metal; and, worst of all, the key to the strongbox was kept in an unlocked desk close by. Two young employees who might have been able to obtain a key to the cage were the first suspects, but a search of their living quarters failed to turn up anything.

Things drifted for over a month with no new suspects despite Bralorne's offer of a thousand dollar reward. The break came when a boy playing in the basement of his home found a rotten canvas sack containing several pounds of high-grade stuffed behind the plank wall. The house, rented by the boy's father, "Porkchop" Campbell, belonged to Gregor McGregor, commonly known as "Truthful" McGregor.

This high-grade had nothing to do with the missing gold sponge, but after discussing it with a senior mine official, a police constable on temporary duty at Bralorne decided to search both the house and Truthful's present home. News travels fast in a mining camp, and Truthful was waiting when the constable left the station to have the just-prepared search warrant signed by the magistrate. Truthful wanted to know more about the high-grade at his old house, but was told that the boy had found some worthless rocks, nothing to get excited over. Shaking off Truthful, the constable rushed to get the warrant signed by the local magistrate, who just happened to be Bralorne's amalgamator.[5]

The search of Truthful's garage turned up equipment used to recover gold from high-grade, plus tools and wire "borrowed" from a hydroelectric project where some of the Bralorne men had worked during the strike a few years before. In addition, there were signs of digging in the yard, most of it supposedly from recent work on the septic tank system. After checking all the disturbed spots the constable was about to give up, when, poking in a freshly-dug flower border at the back of the house, he felt something unusual and dug out a home-sealed tin can. Opened, it contained half of one of the missing segments of gold sponge, and more digging in other spots turned up a total of 14 cans containing all the missing sponge. As reported, Truthful's comment on the find was, "Well... How in hell did that get here!"[6]

Truthful, in the Lillooet jail for 2½ months awaiting a preliminary hearing, had ample time to think up an explanation for it all. He told the court that he had taken his wife and 15-year-old son for a drive on Easter Sunday and that on returning after dark the car's lights had failed just as he turned into his yard. The others went inside while he worked on the car and soon the lights flashed on, revealing "an itinerant wanderer going along the road with knapsack and suitcase" who immediately dropped both and ran off. Truthful, on examining them became suspicious of the weight and, on opening them, discovered the gold sponge. Afraid he would be accused of theft he had sealed the sponge in cans and buried it in his yard. Unconvinced, the magistrate committed Truthful for trial at the Fall Assizes in Quesnel.[7]

Things looked bad for Truthful. Mr. Justice Manson, the assize judge, was known as an unpredictable man apt to hand down stiff sentences, and

Angelo Branca, the defence counsel, had often clashed with Manson in the past. But this trial was different. At some point the judge and lawyers for both sides agreed to plea bargaining, whereby Truthful pleaded guilty to illegal possession but not to the theft of the gold sponge and was sentenced to a mere nine months in jail and a fine of $250.[8] Then, hustled back to Lillooet, he was brought before the magistrate, found guilty of two minor charges of theft and sentenced to an additional 30 days.

No reason was ever given for the plea bargaining. Possibly Bralorne simply wanted their gold back in a hurry, or perhaps the judge was touched by the plight of a fellow Scot with a name like Gregor McGregor. The question of how Truthful got the sponge was never answered. A stope miner, he had no occasion to be around the mill and accomplices, if he had any, were never brought to trial.

At Bralorne there was much speculation over the light sentence. Did Truthful have some sort of deal with the company? One rumour had it that he moved freely underground and, using a set of portable scales, bought high-grade from the miners at $10 an ounce and resold it to the company for perhaps double that. A second had him shipping high-grade to a relative in the Cariboo who then "mined" it on his placer claim. Still another had Truthful paying $5,000 for the sponge after the initial suspects retrieved it from an outhouse where they had buried it in anticipation of a police search.

About six months later, Don Matheson was surprised to see Truthful walking towards the Bralorne office and, thinking quickly, told the mine geologist to phone him on some pretext after Truthful had been with him for about five minutes. It was no mere social call; Truthful wanted his old job back and, on being turned down, broke off the meeting with, "You mean it's the end of my job! My God, Matheson, I thought you were a friend of mine!"[9]

## *Mrs. Noel*

Delina Noel still made her summer home at the cabin on the Lorne property and, over the years, she had become an experienced prospector as well as a claim staker. Now in her mid-sixties, she was still prepared to shoulder a pack and head for the hills. In 1945 she discovered and staked an interesting tungsten-copper showing.[10] Her Chalco property, located on Piebiter Creek about six miles southeast of Pioneer mine, was not easy to reach. A rough road ran along the north side of Cadwallader Creek to the mouth of Piebiter Creek; from there a steep trail led to the campsite on Piebiter at an elevation of about 5,000 feet. Each spring, Mrs. Noel would start moving her outfit to the property as soon as the weather warmed enough to suggest a reasonable chance of getting through. Usually, Pioneer mine provided a truck to move her gear over the passable section of the road and, from there, it had to be carried on the backs of men or horses. Finally, in 1955, ten years after the property was staked, a rough bulldozer road was pushed part way up Piebiter Creek.

Mrs. Noel did some things differently. For one, she refused to ration her food, her attitude being, "When I'm out of food I'll be hungry but I'm not going to go hungry when I have it."[11] Part of her outfit was a pair of breeches, double thickness in the seat, and on descending a steep slope she would often sit down and slide. Even so, the reinforced seat seemed outlast the knees and she was forever patching the latter.

She had been quite a hunter in her early married years and the hide of a near-record grizzly bear she had bagged was mounted on the wall of the Lorne cabin. But now, on her Piebiter ground, she was at peace with the grizzlies and, at times, as many as three shared the valley with her. She recognized them all by sight and enjoyed watching their antics as they fed and played on the open slopes above her camp.

The tungsten-copper showings on her Chalco property were on the bluffs forming the valley wall, close to a thousand feet above the creek level. Work done by Mrs. Noel and her helpers over a ten-year period included locating a number of showings spread over a distance of about a mile and exposing as much bedrock as possible around them, either by hand trenching using pick and shovel and the occasional stick of dynamite, or by ground sluicing. In addition, trails were built to the better showings, something essential as prospecting for scheelite, the tungsten mineral involved, is best done in darkness using an ultraviolet lamp under which the mineral glows a distinctive bluish-white. The alternative, attempting to prospect in daylight while crouched under a black cloth, is most uncomfortable and far from satisfactory.

Filled with a prospector's optimism, Mrs. Noel was certain that her Chalco property would become a major tungsten mine. In 1955, five years before her death in her 79th year, she and one employee were still at it; spending an entire summer prospecting, stripping the showings and building trails.[12] Perhaps her optimism will be justified some time in the future but, until now, neither her work nor later work by a major mining company have come close to proving the tonnage needed for a mine.[13]

## A Road Connection

In the post-war years, the number of cars in the Bridge River valley increased dramatically and so did the demand for a direct link to the provincial highway system. Until now, cars had been ferried across the 15 mile Lillooet-to-Shalalth gap, first by boat on Seton Lake and, in later years, by rail. The PGE's gas-car, towing flatcars laden with autos, made two round trips a day. The need for a road link became even more urgent when that service, already overtaxed, was cut to one trip a day and put on a first-come, first-served basis. Even with the former reservation system, missing the gas-car due to road or car troubles often meant an overnight stay at Shalalth or Lillooet.

Building a road through the canyon of the Bridge River had long been considered impractical but, if a direct link was wanted, there was really no alternative. Heavy rock work would be involved in some sections; elsewhere the steep valley walls bore the scars of recent slides. The hazards could never be eliminated, but all things considered, they were probably no worse than those faced on the switchbacks between Shalalth and Mission Pass.

In the fall of 1954, a road committee representing both business and individuals was formed to lobby the provincial government. One thing was certain: nothing would happen until Phil Gagliardi, the colourful Minister of Highways, was sold on the scheme, and delegations from all

The PGE gas-car on its final run from Lillooet to Shalalth, October 1, 1960. Photo by K. Anderson.

over the province were pestering him for help in one form or another at that time. Ernie Shepherd of B.R.X. Consolidated Mines, given the selling job, pulled it off by catching Gagliardi at lunch and convincing him to attend a meeting in the Bridge River. At the October meeting, the Bralorne Community Hall was packed with close to 500 people, but it was the sight outside that impressed Gagliardi:

> I went to Bralorne via Shalalth, and when I got there you never saw such an assembly of automobiles in all your life; it was simply astounding - I thought they must have sent to Vancouver and hired a fleet of cars for the occasion.
>
> So I finally said to myself if the government is collecting gas and oil taxes on all these cars up here then they should really have a road - so we agreed to go along.[14]

Gagliardi made no commitments but acknowledged the problem and assured the meeting that he would do his best to come up with the solution. His offer, made soon after, was that, if the community provided the men and equipment, work could begin immediately under the supervision of a government engineer. The government, for its part, would pay actual expenses and wages with the proviso that no overtime wages would be paid on the project.

Gagliardi's offer was taken up and work started in late October 1954. It quickly became a community project involving the committee, the men on the job, and even the local ladies who kept the road camp well supplied with home baking. Much of the equipment and personnel came from Bralorne and Pioneer mines and, for once, the two forgot about being standoffish and pitched in to get the job done. The route, just over ten miles long, followed the west side of the Bridge River downstream through the canyon and, at the lower end, crossed the Yalakom River to join the existing road from Lillooet. The weather co-operated and the crew, working long hours and a seven-day week, stuck with it until early February 1955. Work started up again in the spring and there was an unofficial opening on Saturday, September 24, 1955, when a contingent drove through from Lillooet. At a dance held that evening, the big hit was the burning of a replica of the gas-car.[15]

The official opening three weeks later was Gagliardi's day, and he made the most of it. In his opening remarks he praised everyone involved in the project and noted that the government's initial pledge of $15,000 had now grown to $150,000 but, with that outlay, a good 25-foot road had been completed. After Gagliardi cut the red and white ribbons, a cavalcade of 40 cars set out down the new road. With a police car in the lead - perhaps the best if not the only way, of holding "Flying Phil" close to the speed limit - it was on to Lillooet for a Board of Trade lunch and then to Kamloops, Gagliardi's home base, for a Board of Trade dinner and dance.[16]

Building the road had united the community as never before but, with the job done, a letdown followed. The feeling of isolation that had bound the Bridge River's residents together and made the valley unique had ended now that it was just a few hours' drive to the bright lights.

## *Bralorne Loses Don Matheson*

On November 7, 1957, the Bralorne community was shocked by news of Don Matheson's death in Vancouver General Hospital. Taken ill less than a week before, he had been flown to Vancouver and following a major operation, he appeared to be doing well until a sudden relapse and his death that same evening, despite emergency surgery.[17]

To many, Matheson, 52 years old and general manager since 1940, represented Bralorne Mines and all it stood for. Austin Taylor might control the company but he was an austere figure, seldom seen around the mine, and Matheson was the one they dealt with on a day-to-day basis. Well liked, he was always willing to listen and, most important of all, was perceived as being fair. Readily approachable, he sometimes ended up dealing with things that could well have been handled by others, such as a wife's complaint about a company house, but if this bothered him, he never let on. No pushover, he could be tough when necessary, although he did have a soft spot for company employees who had made their contribution in the past and were no longer capable of pulling their full weight. For them, jobs they could still handle were often found or created. He was not replaced immediately and, in the interim, his duties were divided between M.M. "Mel" O'Brien, managing director of the company based in Vancouver, and Cy Manning, general superintendent on the property.

# *Franc Joubin Takes Charge, 1958–1963*

Austin Taylor always played a lone hand, and in the summer of 1958 he quite unexpectedly agreed to sell his controlling interest in Bralorne Mines to Franc Joubin and associates, who already had an agreement in principle to buy a controlling block of Pioneer shares. The Bralorne deal was something of an afterthought, negotiated while the Joubin group was awaiting an answer to their proposal to acquire control of the Pacific Eastern property, Pioneer's neighbor to the southeast, and to explore the ground at depth by extending the Pioneer workings. Older workings, closer to the surface, had been abandoned in 1947 after three miners lost their lives in a methane* explosion on the 520-foot level.[1] The Pacific Eastern deal was never completed, but in February 1959, the two producing mines were merged as Bralorne Pioneer Mines Limited.

For Joubin, then in his late forties and a wealthy man, it was a return to his 1934 start in the exploration game. In a reversal of roles, the deal was in part done to help out Howard James of Pioneer, who had been kind to him in those beginning years. Joubin's first summer in the Bridge River as a prospector for a mining syndicate had ended in disaster when the promoter disappeared without a trace, leaving bills and wages unpaid. Back the following summer and unable to get an underground job, he persuaded James to let him landscape around the latter's new home. The personal contact paid off and in 1937 Joubin was hired as a prospector/geologist for Pioneer. He spent many of the war years looking for strategic minerals and, following this, was posted to Pioneer's Toronto office on a half-time basis until striking out on his own in May 1949.[2] Interested in uranium since his student days, his big find came in 1953 with the recognition and staking of Ontario's Blind River uranium deposits destined to make Canada one of the world's leading producers. Later, after Britain's Rio Tinto group had taken control of the developing mines in return for providing financing, Joubin had joined them as a consultant and, in addition, set up an independent practice.

When the mines merged, Bralorne was the continuing company and its share capital was increased from 1.25 million to 2 million shares. Pioneer shareholders received one Bralorne share for each five Pioneer shares, to

Geologists Franc Joubin (left) and Jim Dawson about 1984 in the field near McGillivray Pass, where the geology has similarities with that at the mines. Photo courtesy of L. Wolfin.

hold not quite 22 percent of Bralorne Pioneer Mines Limited, the ratio struck through an appraisal of the two companies by a firm of consulting engineers.[3] At the time both had close to $1.5 million in net current assets. The big difference was in ore reserves, Bralorne's estimated at 732,000 tons grading 0.70 ounces of gold per ton and Pioneer's a mere 139,280 tons grading 0.48 ounces. For Pioneer that meant less than two years operation, while Bralorne was assured of at least five and appeared to have a far better chance of developing additional reserves at depth on their 77 Vein.

Bralorne and Pioneer were now one in the corporate sense, but combining the two into a single, lower cost, and more efficient operation was another matter. Although less than three miles apart, the two operations had gone their separate ways from the beginning with surprisingly little communication and co-operation between them. For the moment, they continued to operate independently, with staff apprehensive of the changes that would soon be upon them.

A few months later, two senior mining engineers with no ties to either operation were added to management. J.E. McMynn, 44, with extensive experience in western Canada, spent six months in the Bridge River and in August 1959 was appointed general manager of Bralorne Pioneer, based in Vancouver. About the same time, C.M. Campbell, Jr., 45, with a background in both gold mining and exploration, was appointed resident manager at Bralorne. Campbell was the man on the spot when the decision was made to shut down the now-marginal Pioneer operation in

Bralorne from the south in the early 1960s, with the new Bralorne-Pioneer ski cabin under construction in the foreground. Bradian townsite is in the middle ground behind the ski cabin. The main Bralorne workings and town are to the left. Cadwallader Creek flows from right to left in front of the town and mine; the Bridge River occupies the large valley running across the photo about a third of the way down. Photo courtesy of C. Campbell.

Bralorne staff with gold brick No. 3,000, poured September 29, 1963. The men are grouped around a wooden frame representing the total volume of gold produced to date. From left to right: A.J. "Jack" Learmonth, plant superintendent; C.M. Campbell, Jr., resident manager; Jack Marshall; J.P. "Peter" Weeks, chief geologist; Earl Weston, mill foreman; W.E. "Bill" Field; unkown; Ed Hall, mill superintendent; Peter Wiseman, chief accountant; and J.S. "Jim" Thomson, mine superintendent. Photo courtesy of C. Campbell.

August 1960. Some older members of the Pioneer crew retired; others chose to leave the Bridge River; but Campbell was able to fit almost all those who wished to stay on into his Bralorne operation.

Before the shutdown, Pioneer's No. 5 Shaft had been sunk to 30 Level and the 27 Vein drifted on it. In addition, a diamond drill program tested the Main and 27 Veins, 1,000 and 1,500 feet respectively, below 29 Level. Both were there; the Main, narrow with low gold values, and the 27, narrow to very wide with low to spotty high gold values, the highs confined to a small section not worth going after. Bralorne and Pioneer had not been joined underground and the shutdown left a block of favourable ground, roughly 2,000 feet across, unexplored except for a few drill holes. Pioneer still had an ore reserve of 64,000 tons carrying 0.46 ounces of gold per ton and, for the moment, its mine and mill were kept intact.

Bralorne mine, in contrast to Pioneer, was doing well and appeared to have a future. For this, if no other reason, it made good sense to shut Pioneer down and concentrate on whipping the Bralorne operation into shape. By 1960 results were showing: gold production for the Bralorne division in that year was 114,115 ounces, highest in the mine's history, and the cost of producing each ounce had dropped to $25.62, down from $29.28 the preceding year. Ore reserves were down very slightly and profits for the year of 51¢ a share more than covered the 40¢ in dividends. The internal Queen Shaft was now down to 39 Level and the highly productive Coronation (77) Vein would soon be tested on 38 and 39 levels.

In 1961 a new 600-ton-per-day mill costing close to $500,000 replaced the old Bralorne mill in use since 1931. The building and most of the machinery were brand new. A modern cyanide circuit replaced the blanket tables and flotation plant, doing away with the need to ship a concentrate and all the ensuing problems. There was a link with the past, too, in the continuing use of jigs and in the amalgamation of the jig concentrates to recover about 65 percent of the gold produced.[4]

By 1962, Joubin's interests were turning elsewhere and he gradually withdrew from the Bralorne project. Late in 1963, George H. Davenport was named president and chief executive officer. Davenport had joined the accounting department at the mines in 1933 and since 1955 had served as secretary-treasurer and as a director of the company.

In 1971, forty years after the first Bralorne school opened, teacher Margaret Pasacreta watches Wayne Keir ring the Bralorne school bell for the last time. Photo courtesy of Margaret Pasacreta.

# *The Final Years, 1963–1971*

By 1965, when Joubin had disposed of his Bralorne Pioneer shares and left the board, the Bralorne mine faced an uncertain future. It had not been for lack of trying; during the Joubin regime major improvements had been made at the mine. Despite increases in miners' wages and the problems of mining at increasing depths, a better than 25 percent increase in productivity had held costs per ton of ore to a mere 7 percent above the 1957 figure. However, fate, or "Miss Serendipity" as Joubin sometimes called her, had looked the other way and not one of the three things that could have made the difference had come to pass. The price of gold was still stuck at $35 US an ounce; no important finds of high-grade ore brightened the ore reserve picture; and outside exploration had failed to come up with a new mine or mines to replace Bralorne.

Under Davenport's leadership the company began to turn away from mining and, after acquiring substantial oil and gas interests, incorporated a subsidiary company to manage them. In another, totally unrelated, venture, the company invested more than $1 million in a subsidiary company manufacturing prestressed concrete products. As for the Bralorne mine, for five more years it operated at close to 300 tons per day with gold production close to the 50,000 ounces a year mark. That was nothing to get excited over; the cost of producing each ounce of gold had passed the $35 US an ounce mark in 1965 and the only profit was a portion of the government aid paid out under the Emergency Gold Mining Assistance Act. In February 1966, Bralorne Pioneer's directors recognized the inevitable and suspended dividends indefinitely after a final payment of 5¢ a share.

The Bralorne mine continued to operate, perhaps in hope that the long-anticipated increase in the price of gold could not be put off much longer. There was still some exploration and, in July and August 1965, production was curtailed while the Queen Shaft was sunk an additional 373 feet from 41 to 43 Level, the latter only a few tens of feet shy of a mile below the main haulage on 8 Level.[1]

During 1967 development work on 42 and 43 levels added substantially to the mine's ore reserves which, at the year end, were 170,900 tons grading

0.53 ounces of gold a ton, up almost 40,000 tons from the year earlier despite mining of 97,000 tons. Now, instead of the 1968 shutdown predicted at the end of 1966, the company was considering sinking below 43 Level.

Working on 43 Level was a challenge. Rock temperatures, which increase with depth, were 128°F (53°C). Ventilation was increased using two 150-horsepower fans on the surface to force 100,000 cubic feet of air per minute down a 12-foot diameter ventilation raise to 25 Level and thence down the Queen Shaft to the working levels. From there, auxiliary fans forced the air out to the working faces through specially designed ventilation pipes covered with an inch of fibreglass insulation to keep the air as cool as possible. From the working faces the exhaust air rose through the stopes and other workings on the veins to 26 Level and thence by way of the Crown and Empire shafts to the surface. An additional barrier to the heat was provided by a layer of insulating urethane foam sprayed on the back and wall of the crosscuts and drifts.[2]

The worst problem was in driving raises where, with exhaust air unable to escape upwards, temperatures rose to intolerable levels. Raises were an essential part of the mining plan, and Bralorne's solution was to replace miners by a raise boring machine. In its operation, a 9-inch diameter pilot hole was drilled from an upper level to a lower where the drill bit was replaced with a 48-inch diameter reaming head and the hole reamed upwards at the full width. In 1966, a total of 1,404 feet of raise-boring was done with a 345-foot raise the longest of five completed.[3]

Barring an increase in the price of gold, mining the new reserves would do little to improve Bralorne Pioneer's financial picture. Gold mining was marginal at best; the company's oil and gas operations provided a modest return, while the venture into concrete technology appeared to be going nowhere. Despite all this, others appeared to sense an opportunity in Bralorne Pioneer. During 1967, two bids were made for the 340,000-odd unissued shares remaining in the company's treasury. Both were declined but Can-Fer Mines Limited, American-backed with assets in the $1 million range, began accumulating shares in the open market, eventually acquiring 250,000. Early in 1968, two of their officers joined the Bralorne Pioneer board. In 1969 the two companies were merged to form Bralorne Can-Fer Resources Limited. Bralorne Pioneer shareholders, either outmaneuvered or stung by events of the past few years, were left with less than a 30 percent interest in the continuing company.[4]

During 1968 and early 1969, a series of declines* were sunk on the eastern extension of the Coronation (77) Vein to test this remarkable structure on 44 and 45 levels. The ore continued and, at the end of 1969, reserve figures were essentially unchanged from those of two years before.

Despite operating problems, mining continued in 1970 as returns, including government payments under Emergency Gold Mining Assistance Act of more than $10 for each ounce of gold produced, still supplied a modest cash flow and tax credits that could be used in Bralorne Can-Fer's other ventures. The change was that a once-great mine had now turned into a salvage operation destined to end as soon as gold could not be produced at an acceptable cost. There was no planning for tomorrow; attempts to find new ore reserves to replace those being mined were abandoned and operating expenses cut to the bone.

By now there were fewer than 150 names on the payroll, a gradual decline from 300 in 1964 and earlier years. A few of those leaving had been old-timers retiring, others had moved to different mines, but many had left the mining game for good, convinced that underground mining was on the way out. It was a time of change. Mining in the province was booming, but much of the production was from open pit operations treating vast tonnages of low-grade ores of copper and molybdenum. In them, rising labour costs could be offset using bigger, more efficient machinery to break the cost spiral that threatened to put many underground mines out of business. For most of the sixty-odd employees still around in the summer of 1971, there was a reprieve of sorts in an offer of jobs at the Bradina project, near Houston, B.C., where a silver-zinc-copper property was being put into production as an underground mine. Moving expenses of those who signed on would be shared between Bralorne Can-Fer, which held a 25 percent interest in the joint venture, and Canada Manpower, a government agency.[5]

Those staying on witnessed the collapse of their once-vibrant community. Company housing was spread out in a number of sites, making it difficult to patrol. All too often, vacated homes were vandalized within 24 hours. No one seemed to know who the culprits were and perhaps it was of no consequence, as one rumour had it that the company, responsible for cleaning up the site, planned to salvage what they could, then burn the houses and let nature take over.

Mining finished at the end of August 1971, milling and clean-up during the next month. As 1971 drew to a close two caretakers were the only residents remaining on the company payroll.[6] In total, the Bralorne property had produced 2,821,567 ounces of gold and 706,345 ounces of silver from the 5,461,400 tons of ore milled and Pioneer 1,333,074 ounces of gold and 244,648 ounces of silver from 2,469,700 tons of ore milled.[7] Total production for the Bridge River camp is 4,178,363 ounces of gold, including 17,557 and 5,341 ounces from Minto and Wayside properties, respectively, making it the largest lode gold producer in B.C. and a close second to the total estimated B.C. placer production of 5.3 million ounces.

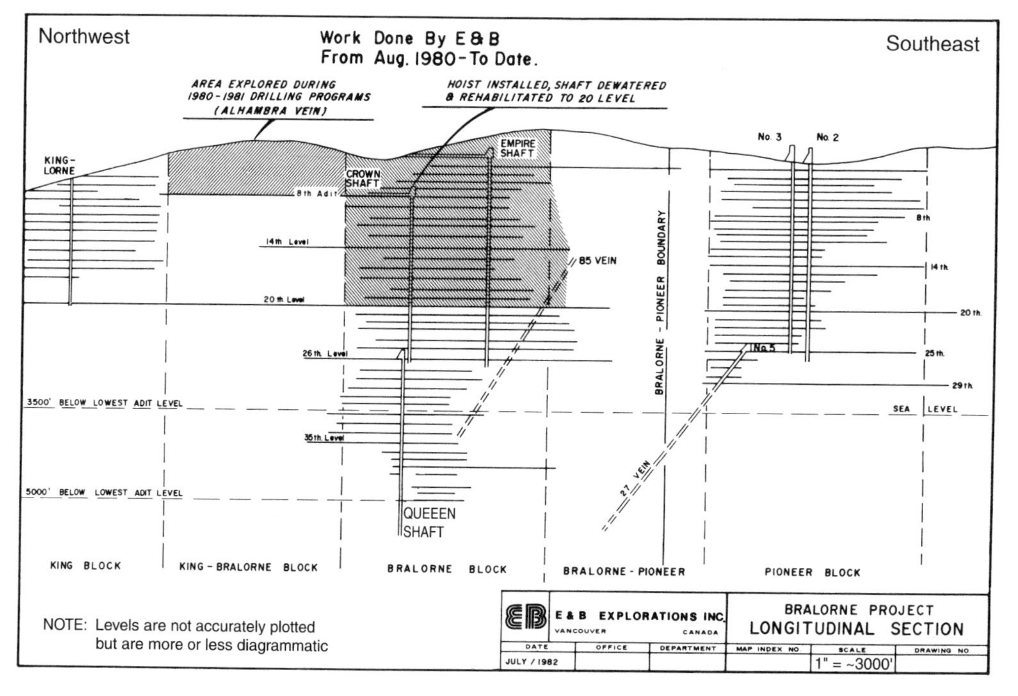


A diagrammatic cross-section through the Bralorne and Pioneer workings with the distance between the King-Lorne and Pioneer No. 2 shafts being roughly 12,000 feet. The horizontal lines are levels, the horizontal passages used for mining. The levels are connected with each other and with the surface by vertical shafts, shown by thin double lines. The shaded areas were explored by E&B Explorations during the early 1980s, illustrating that even heavily mined areas may be worth another look. The diagram also shows that there is much untested ground between Bralorne and Pioneer that may (or may not) contain ore. Diagram courtesy of H. Ewanchuck.

# *Afterword*

Since the Bralorne closure, exploration aimed at reworking the Bralorne ground or finding new deposits has gone on by fits and starts but, as yet, not another ounce of lode gold has been produced. Ironically, mining ended just over two weeks after August 15, 1971, the day President Nixon abandoned the fixed gold price of $35 US an ounce by suspending convertibility of the United States dollar to the metal. Gold began to rise slowly but erratically, hitting $100 US an ounce in 1973 and on January 21, 1980, shooting up to an all-time high of $850 US at a time of near-panic in financial markets. Falling back the following year it was mainly in the $350 to $450 US range until late 1997 when it dropped below the $300 mark. Even at $450, more than 12 times the $35 US price of 1934, it is no bonanza when compared to a more than twenty-fold increase in the cost of living since 1945.[1]

Worldwide, gold production continues to rise, with much of it coming from large, low cost, open-cut* operations where gold is recovered by heap leaching with a cyanide solution. In these mines, increasing labour costs can be offset by the use of larger and more efficient machinery. Not so in underground gold mines such as Bralorne, where following narrow veins remains labour-intensive and, with skilled miners hard to come by, costs have risen faster than the inflation rate.

The gold bugs are still at it, most predicting increases in the price of gold and a few preaching financial meltdown with the purchasing power of gold rising to astronomical levels. In a world awash in both paper and paperless money, all with little or no real backing, and the increasing complexity of financial transactions, who can foretell what lies ahead? A return to a gold standard seems unlikely with central banks in a number of countries selling a portion of their reserves and, in the process depressing the price of gold to a 20-year low of $252.55 US on August 25, 1999. Just over a month later the price bounced back above the $300 US level after the European community agreed to limit sales of gold reserves to no more than 400 tons a year over the next five years. For gold producers, the best of all possible worlds might well be to have the metal simply a commodity with a price determined in the marketplace, free of meddlesome central banks.

The Bralorne company, renamed Bralorne Resources Limited in 1972, moved to Calgary, Alberta, where it was involved in the oil and gas industry, both in exploration and production and in the supply of equipment and services. After 56 years it no longer holds a direct interest in the Bralorne mine. In August 1980, it entered into an agreement giving another company the right to explore the ground and then, in 1987, their remaining interest was sold to a third party for $2.5 million and a small royalty on the net proceeds of any future production.[2]

Shortly after the mine closed, surface rights to all but 133 acres of Bralorne's ground were acquired by the three Whiting brothers of Vancouver, operating as Marmot Enterprises. Their objective was to develop a community serving as a retirement spot for seniors, a summer vacation area for two-home families, and a winter resort. Initially, only Bralorne's No. 1 and No. 2 townsites were involved, as most houses in the Bradian townsite were either beyond repair or close to it, as were the few remaining at Pioneer after most had been burned. The Bralorne water and sewer system was largely replaced or upgraded by Marmot, and many of the hundred-odd houses remaining in Nos. 1 and 2 townsites were sold. Subsequently, a new gravelled switchback road was built to service 45 one-acre lots developed on the hillside above the Bradian townsite. The enterprise, hit by recession and government red tape, was a commercial failure but a social success inasmuch as Bralorne survives as a residential and vacation community.[3]

For a brief period in the late 1970s, the Bralorne Community Club attempted a commercial venture, growing mushrooms in the mine's 8 Level adit. The product was of excellent quality, but the scheme was abandoned when failure to use sterilized manure brought on a plague of flies whose larvae filled the mushrooms with tiny worms.

In 1995, 24 years after the closure, preparations were underway to reopen the Bralorne mine. The project was a joint venture by International Avino Mines Ltd. and Bralorne-Pioneer Gold Mines Ltd., both with Louis Wolfin as their president. Avino had acquired the mineral rights to the Bralorne and Pioneer ground, plus a remaining minority interest in the adjacent Loco property, formerly Taylor (Bridge River) Mines in 1991, and in 1993 optioned a half interest to Bralorne-Pioneer Gold.[4] During 1995 the British Columbia government issued the required Mine Development Certificate and preliminary financing was negotiated. Initially, ore was to be mined from above 8 Level in both the King and Bralorne mines where the companies reported "readily available" mineable reserves of 476,835 tons grading 0.31 oz gold per ton.[5]

However, like Austin Taylor's 1931 gamble, long-term success hinges on the discovery of new reserves. These may well be present on the Loco property northeast of the original Bralorne ground. Here, new showings, consisting of banded quartz veins similar to and carrying gold values comparable to the productive veins, were discovered in a program of detailed geochemical prospecting directed by geologist Jim Miller-Tait. The finds have been exposed and sampled in surface trenches and one, the Peter Vein, explored from a short adit where a 215-foot section assayed 0.38 ounces of gold per ton over a 3.4-foot average width.[6]

Both the Peter and a second vein, the Millchuk, lie on the Cosmopolitan claim (staked August 28, 1897, and later Crown granted) and have been traced into the adjoining Noelton claim. The latter may yet prove to be worth far more than the $65,000 Mrs. Noel got for it in 1933.

Milling equipment purchased early in 1996 has been installed in the reconditioned mill building and could be put into production quickly following construction of a tailings pond. However, further work on the project has now been deferred pending an increase in the price of gold and more favourable financial conditions.

Travelling the road from Bralorne to Gold Bridge, all that marks the site of Ogden is a widening in the road and some twisted metal, the remains of the Mines Hotel, destroyed by fire in June 1984. Zada's sporting house and the other buildings have disappeared without a trace. Farther along the road, Brexton, first known as Fish Lake and never very large, still has a few buildings, some old, some new.

Gold Bridge, with a general store, post office, garage, hotel, motel and highways yard, is still the centre for the Bridge River's residents. Its three earlier hotels, the Goldbridge, the Royal and the Truax, the latter cut in two and moved down from Brexton in the 1960s, are all gone, destroyed by fire. The place can still kick up its heels a bit on Saturday night. But at dusk on a weekday evening in early June the ghosts are quiet, the air soft following a light shower and filled with the perfume of cottonwood buds, while swallows, seemingly the only thing about, swoop within a few feet of the ground. Scattered homes, old and new, all have their satellite dishes aimed skyward. In the graveyard, characters from the past, like the self-proclaimed King of Norway, have found their last resting place. Half a mile away at Haylmore, most of the buildings and rock walls are gone, bulldozed away in the search for placer gold.

Between 1946 and 1960, much of the Bridge River's valley floor disappeared beneath the waters of two artificial lakes created to store water for the B.C. Electric Company's Bridge River power development. Just over a mile above Gold Bridge, the Lajoie Dam holds back the waters of Downton Lake, some 16 miles long, and a small 22,000-kilowatt plant supplies electricity to the local area. Below Gold Bridge, Carpenter Lake, over 30 miles long, has been created behind the Terzaghi Dam built across the mouth of the Bridge River canyon, flooding out the meandering river flats where Arthur Noel's crew poled whipsawn boats laden with the Bend 'Or machinery in the spring of 1899. Tunnels, the first holed through in 1930 and the second in 1948, carry silt-laden Bridge River water beneath Mission Ridge to penstocks of power plants on the shore of Seton Lake.

Leaving Gold Bridge, the road crosses the river and a branch to the left leads up the hill to Lajoie Lake, still Little Gun Lake to the old-timers, where the lodge built for Pioneer's three "Ss" still stands, a bit weathered but otherwise as impressive as it was the day Big Bill Davidson completed it in 1933. Sold in 1940, the lodge has had a succession of owners since and, most recently, was operated as a guest lodge between 1981 and 1986. Beyond, the shores of Gun Lake are crowded with cottages, some now in use as year-round residences.

The main road to Lillooet squirms along the hillside a few tens of feet above the high-water mark of Carpenter Lake. For the first five miles, signs of recent mining exploration are visible above and below the road and, at times of low water, old foundations, waterlogged timbers and waste dumps appear on the lake bed. Beyond Gun Creek the outline of Big Bill Davidson's Minto townsite is visible at extreme low water. Following the departure of the Japanese after World War II, Minto boomed briefly when men working on the B.C. Electric power development brought their families here. There was nothing going on at Davidson's Minto mine but he was still in town and, when the weather was suitable, spent much of his time on the porch of the Minto Hotel playing crib with a crony. First one and then the other would win and an incredibly dirty dollar bill was passed back and forth between them. The end came in June 1950 when a flash flood on Gun Creek hit the townsite, leaving behind a trail of boulders and debris. The town, nearly empty at the time, was never rebuilt and Davidson retreated to the Minto mine's staff house on the hillside above, his home until his death in the Bralorne hospital on January 15, 1958, at the age of 73.

Thirty-six miles out from Gold Bridge, the Terzaghi Dam holds back the waters of Carpenter Lake and a bronze plaque honours its namesake, the founding father of modern soil mechanics. From here, Lillooet is a mere 15 miles to the southeast as the crow flies but twice that distance by road, first through the canyon where the crew of local residents drilled, blasted and cleared a roadbed in 1954–55 and, after crossing Yalakom

River, along an older section of road that follows a narrow bench on the Bridge River valley's north wall.

Crossing on the Terzaghi Dam one can still travel the old route over Mission Pass to Shalalth, although it will involve backtracking unless one is adventurous enough to tackle the rough access road that follows the powerlines above Anderson Lake and rejoins the provincial road network at D'Arcy. After a three-mile stretch along the south shore of Carpenter Lake the road switchbacks up to the pass through a logged-over area where traces of the earlier route have been obliterated. Beyond the summit, one looks out on the dull turquoise waters of Seton Lake, coloured by silt and, to the southwest, on the deep blue waters of Anderson Lake. The road lacks guardrails but has a good gravel surface and adequate turning room for a car on the corners of the switchbacks zigzagging down the 3,000-foot descent to Shalalth. Signs warn that trucks in excess of 45 feet long are prohibited. Shuttling back and forth on the steep hillside soon becomes boring, but one should try and visualize what it must have been like when the road was narrow and cut up by heavy traffic. Then, drivers had to contend with traffic jams, shifting loads, blowouts and mechanical failures plus the seasonal hazards, ice and snow in winter and spring, excessive heat in summer and mud following a heavy rain or a sudden thaw.

At the bottom, a simple shed now serves as B.C. Rail's Shalalth station. To the east of it lie scattered homes on the Indian reserve and, to the west, the power plants, now part of the B.C. Hydro system, and employee homes. Everything is quiet now, the hustle and bustle of the 1930s forgotten. Then it was one of the busiest stations on the PGE line with sidings, freight sheds and transportation company yards that, each year, saw tens of thousands of tons of freight and thousands of people dispatched to the booming mines of the Bridge River.

# *Glossary*

Most of the following definitions were modified from *A Dictionary of Mining, Mineral and Related Terms* (see bibliography).

**Adit** A nearly horizontal tunnel driven from the surface for the working of a mine.

**Amalgam** The pasty mixture of gold plus minor silver with mercury, about one-third gold by weight.

**Amalgamation** The production of an amalgam.

**Arrastra** A circular, rock-lined pit in which broken ore is pulverized by stones attached to horizontal poles fastened in a central pillar and dragged around the pit.

**Arsenopyrite** A tin-white, sulpharsenide of iron, FeAsS.

**Back** The roof or upper part of any underground mining cavity.

**Backfill** Waste rock, mine tailings or surface material run into a mined-out stope to reduce subsidence and caving.

**Ball mill** A piece of milling equipment used to grind ore into small particles. It is a cylindrical shaped steel container filled with steel balls into which crushed ore is fed. The ball mill is rotated, causing the balls themselves to rotate, which in turn grinds the ore.

**Bar down** To test for loose rock in the back or walls of a working using a heavy steel rod with a pointed or flattened end and to dislodge it.

**Blanket table** A sloping table covered with corduroy blanket material or cocoa matting over which finely ground ore is washed to recover free gold and associated minerals such as pyrite and arsenopyrite. The blanket material is taken up periodically and washed in a tub to remove a concentrate from which gold is recovered by amalgamation.

**Blind vein** A vein that does not extend to surface.

**Brattice** A barrier of lumber and cloth used to control the flow of air underground.

**Bullion** Gold and contained silver in the form of bars.

**Cage** The structure in a mine shaft used to convey men and materials.

**Carbide lamp** A miner's lamp utilizing an open flame of burning acetylene gas generated by water dripping on calcium carbide.

**Chute** An inclined opening, usually constructed of timber and equipped with a gate, through which ore is drawn from a stope into mine cars on the underlying level.

**Cleanup** Halting the operation of a placer mine, arrastra or stamp mill to collect the accumulated amalgam or precious metals.

**Concentrate** A product containing the valuable metal and from which most of the waste material in the ore has been eliminated.

**Contact** Bounding surface between two rock units, especially the boundary between an intrusive and its host-rock.

**Cousin Jack** A Cornish miner, usually far from home.

**Crosscut** A horizontal opening driven across the course of a vein, especially a connection from a shaft to a vein. Compare **drift.**

**Crown granted mineral claim** A mineral claim on which work and certain other requirements have been met and on application and payment of fees a Crown grant issued, in effect converting the mineral rights on the claim to property dealt with under the provisions of the Land Registry Act.

**Cut-and-fill stoping** A stoping method in which the ore is excavated by successive slices, working upward from a level. After each slice is blasted down, all broken ore is removed and the stope is filled with waste up to within a few feet of the back before the next slice is taken out, leaving just enough room to provide working space.

**Cyanide** Sodium cyanide, NaCN.

**Cyanidation** The process of extracting gold and silver by dissolving and subsequently precipitating the metals from an aerated, weak solution of cyanide in water.

**Decline** An inclined winze suitable for ore haulage.

**Diamond drill** A drilling machine with a rotating, hollow, diamond-studded bit that cuts a circular channel around a core, which can be recovered to provide a more or less continuous sample of the rock penetrated.

**Dilution** The contamination of ore with barren wall rock* in stoping and the resulting lowering of grade.

**Diorite** An igneous* rock composed of an intermediate plagioclase feldspar and a dark mineral, commonly hornblende or pyroxene.

**Dip** The inclination of a bed, vein, fault etc., measured from the horizontal.

**Drift** A horizontal opening driven following the course of a vein. Compare **crosscut.**

**Drifter** A heavy drill, mounted on a column or bar and operated by compressed air, used to drive horizontal workings, both drifts and crosscuts.

**Drive** See **drift.**

**Dry** A miners' changehouse, usually equipped with lockers, showers and a means of drying wet clothing.

**Electrum** A gold-silver alloy with up to 26 percent silver; the "free gold" of the Bridge River is actually electrum.

**Face** In an adit, tunnel or stope, the end at which work is in progress or was last done.

**Fault** A fracture or fracture zone along which there has been displacement of the two sides relative to one another parallel to the fracture. The displacement may be a few inches, hundreds of feet or even miles.

**Float** Loose pieces of ore or rock found on or near the surface or in stream beds.
**Flotation** A method of concentrating ore by inducing particles of gold-bearing sulphide minerals to float to the surface of water when buoyed up by air bubbles while particles of **gangue** or waste sink to the bottom.
**Footwall** The wall or rock on the underside of a vein or fault.
**Gangue** Useless minerals, mostly nonmetallic, occurring in ore.
**Grizzly** A grating, usually of steel rails, placed over the top of a chute or ore pass for the purpose of stopping larger pieces of rock or ore which are then broken by hand or blasted.
**Ground sluicing** Diverting a stream of water across an area to be tested in order to strip away soil, river and glacial deposits, and decomposed bedrock.
**Hanging wall** The wall or rock on the upper side of a vein or fault.
**Hard rock** Rock which requires drilling and blasting for its economical removal.
**Heading** An underground working being driven to a definite objective.
**Heads** The ore fed to a mill. Compare **tails.**
**Heavy ground** Sections underground that are either broken by faults or where the rock is altered to clay and other secondary minerals and workings must be reinforced by timbering or other means to reduce the danger of collapse.
**High-grade** Exceptionally rich ore, which in the Bridge River frequently contained visible gold.
**High-grading** The theft of such ore by miners and others.
**Hydraulic** To move gravels and other surface material by means of a stream of water under high pressure.
**Igneous rock** Rock formed by the solidification of molten material that originated within the earth.
**Jig** Appliance in which a vertically pulsed column of water is so manipulated as to stratify crushed ore with lighter particles above and heavier below.
**Lagging** Small timbers placed to form a ceiling or a wall and to prevent rocks from falling through but not meant to carry the main weight.
**Ledge** An obsolete term for a **vein** or **lode.**
**Level** Mines are customarily worked from shafts through horizontal passages or drifts called levels. Levels in the Pioneer mine were established at 125-foot intervals and in the Bralorne mine at 150-foot intervals and in both were numbered in sequence from the surface down.
**Lode** A mineral deposit in solid rock. **Vein** and, in the past, **ledge** are used in much the same sense.
**Manway** A timbered compartment with ladders in a shaft or stope, men using in the former to move between levels when the hoist is unavailable and in the latter to reach the working face.
**Metamorphic rock** Any rock which has been altered by heat or intense pressure, causing new minerals to be formed and new structures in the rock.
**Methane** A flammable gas, $CH_4$, which can form an explosive mixture with air.
**Mill** A mineral treatment plant in which ore is crushed, ground, and processed.

**Mineral claim** A square or rectangular block of ground with sides up to 1,500 feet staked by the holder of a free miner's licence with intent to explore for valuable minerals in the underlying bedrock. Sometimes referred to as "lode" or "quartz" claims.

**Moose pasture** Derisive term for ground, often covered with muskeg and without known values of any sort, that has been staked near a mine or prospect for promotional purposes.

**Muck** a. Useless material such as barren rock. b. To move waste or ore, usually with a shovel, hence "mucker" and "muckstick" miners' slang for shovel.

**Nipper** A stope miner's helper, often a young man on his first job underground.

**Open-cut** A surface working, open to daylight.

**Ore** In gold mining, rock, often quartz, from which gold and associated precious metals can be recovered at a profit.

**Ore pass** A vertical or inclined passage for the downward transfer of ore, equipped with gates or other appliances for controlling the flow.

**Pan** To wash gravel or crushed rock in a gold pan in the search for gold or other valuable minerals.

**Placer** Deposits of gravel or other surficial material containing gold values, as distinct from lode or hard rock deposits.

**Pulp** A mixture of finely ground ore and water produced in the milling process.

**Pyrite** A brass-yellow iron sulphide mineral, $FeS_2$; it and arsenopyrite are the most common metallic minerals found in the gold-bearing quartz veins of the Bridge River.

**Quartz** A common mineral, silicon dioxide $SiO_2$. The gold-bearing quartz veins of the Bridge River, consisting mainly of milk-white quartz, commonly exhibit ribbon structure due to the presence of small amounts of other minerals.

**Raise** A vertical or inclined underground working driven upward from a level to connect with a level above or to explore ground for a limited distance above a level. Compare **winze.**

**Retorting** Removing mercury from an amalgam by volatilizing it in an iron retort, conducting it away and condensing it.

**Rockfall** The relatively free falling of a newly detached segment of bedrock of any size from a cliff, steep slope, cave or arch.

**Round** A set pattern of holes drilled, loaded with dynamite and blasted each shift in advancing an underground working.

**Sedimentary rock** Rock formed by the accumulation of sediment in water.

**Serpentine** A soft rock composed of hydrous magnesium silicates and commonly green in colour.

**Shaft** A vertical or inclined passageway used to service a mine using a hoist by which men and materials may be raised or lowered. Most shafts are sunk, but in an operating mine a new shaft is sometimes raised from a lower level.

**Shrinkage stoping** Ore is mined in successive horizontal slices working upward from the level. After each slice is blasted down enough broken ore is drawn off from below to provide a working space between the top of the pile of broken ore and the back of the stope. Usually about 40 percent of the broken ore is drawn off when the stope has been mined to the top. The remainder is then drawn down and the stope backfilled with waste rock or other material or, in a few instances, left standing open.

**Silicosis** A debilitating and sometimes fatal disease of the lungs caused by prolonged inhaling of airborne particles of siliceous materials.

**Skip** A self-dumping type of bucket used in a shaft for hoisting ore or rock.

**Sponge gold** A porous mass of gold and minor silver obtained by heating amalgam to distill off the mercury.

**Stamp mill** A machine for crushing ores, used commonly in gold mining but now obsolete. The crushing member or stamp is dropped on a die, the ore is crushed in water between the stamp and die and the crushing space surrounded by a mortar box equipped with a screen to regulate the size of the discharge.

**Steel** Drill steel of varying lengths used in drilling holes for blasting.

**Stope** An excavation in a mine from which ore is being or has been extracted. See **cut-and-fill stoping** and **shrinkage stoping.**

**Stoper** A drill used in stope mining, designed to drill upwards and held in position by a telescoping leg. Compare **drifter.**

**Strike** The bearing of a horizontal line in the plane of an inclined fault, vein or other stratum.

**Sulphide** A compound of sulphur with one or more metallic or semi-metallic elements, e.g., pyrite ($FeS_2$) or sphalerite ($ZnS$).

**Tailings** The waste product from the mill after the extraction of the precious metals. Also known as **tails,** they contain some values that cannot be economically recovered. Constant assays of both head and tails are taken to determine the overall recovery and to control the milling operation.

**Tram** To haul or push ore cars in a mine, either by hand or by a motor.

**Vein** A mineral deposit in solid rock having a more or less regular development in length, width and depth.

**Wall rock** The country rock immediately adjacent to a vein or lode.

**Weathering** The group of processes, such as the chemical action of air and rain water and of plants and bacteria and the mechanical action of changes of temperature, whereby rocks on exposure to the weather change in character, decay and finally crumble into soil.

**Wingdam** A dam designed to deflect water out of a portion of a stream channel to permit placer mining or the underlying gravels.

**Winze** A vertical or inclined opening sunk from a point inside a mine. Compare **raise.**

**Wire gold** Native metal in the form of wires or threads.

**Wood stave pipe** Pressure pipe used to carry water at heads of up to several hundred feet, consisting of specially sawn wood staves fitted together and strapped in place by encircling steel bands.

# Notes

*The Setting*

[1] The Miner, August 1934, 341.
[2] Leitch et al. Geochronometry of the Bridge River Camp, 205.

*Early Lode Mining, 1896–1923*

[1] *British Columbia Minister of Mines Report,* 1886, 207.
[2] Ibid., 1896, 548–549.
[3] Wollacott. "Gold in Lillooet" *Maclean's Magazine,* March 1, 1934, 13.
[4] Ibid., and Edwards. *Short Portage to Lillooet,* 164–165.
[5] *Minister of Mines,* 1897, 554–556.
[6] Ibid.
[7] Ibid.
[8] Ibid., 1898, 1099.
[9] *The Prospector,* November 4, 1898.
[10] Ibid., October 12, 1900.
[11] Ibid., November 18, 1898.
[12] Ibid., June 16, 1899 also see *The Lillooet Prospector,* January 19 and February 2, 1912.
[13] *The Prospector,* June 30, 1899.
[14] *Minister of Mines,* 1910, K142.
[15] *The Prospector,* September 22, 1899.
[16] Ibid., October 13, 1899, and March 23, 1900.
[17] Ibid., March 23, 1900.
[18] Ibid., December 15, 1899.
[19] Ibid., December 29, 1899.
[20] *Minister of Mines,* 1900, 908.
[21] Ibid., 1901, 1093.
[22] *The Lillooet Prospector,* July 4, 1913.
[23] *Minister of Mines* 1913, 262.
[24] *The Lillooet Prospector,* March 29, 1912.

*Pioneer: Dave Sloan Brings in a Mine, 1924–1932*

[1] *The Prospector,* July 14 to October 21, 1898.
[2] Mining Records, Lillooet, *Bill of Sale Record Book 1898–1908,* 62, 93 and 112.
[3] Ibid., 136, December 12, 1902.
[4] *Minister of Mines Report,* 1905, J208.
[5] Ibid., 1910, K143.
[6] *Land Registry,* Kamloops. Record on microfilm "Kamloops DD's" Item 21627.
[7] Mining Records, Lillooet, *Bill of Sale Record Book 1898–1908,* 171.
[8] McCann. *Geology and Mineral Deposits of the Bridge River,* 92.
[9] *Minister of Mines Report,* 1914, K371.

[10] Public Archives of British Columbia: Registrar of Companies GR1526, Microfilm B5193, *Pioneer Gold Mines Limited,* File 2975, 122 entries.
[11] Vancouver City Archives, Pamphlet 1920–18, Pioneer Gold Mines Limited.
[12] *Minister of Mines,* 1918, K229.
[13] Sloan. "History and Early Development of the Pioneer Gold Mine." *The Miner,* August 1934, 339–342.
[14] *Minister of Mines,* 1921, G193, and 1922, N136.
[15] Sloan. "History and Early Development of the Pioneer Gold Mine." *The Miner,* August 1934, 339–42.
[16] Joralemon. *Adventure Beacons,* 300.
[17] Sloan "History."
[18] *Minister of Mines,* 1926, A190.
[19] Ibid., 1928, C216.
[20] *The Financial Post,* March 31, 1934, 32.
[21] *Minister of Mines,* 1929, C231.
[22] Pioneer Gold Mines of B.C. *Chronological Table of Events,* 2.
[23] Eklof interview.
[24] *Minister of Mines,* 1930, A201.
[25] *The Miner,* Pioneer number, August 1934.
[26] *Minister of Mines,* 1932, A222–224.
[27] Ibid., 1933, A265.

***Lorne Gold: An Honest-to-God Promotion, 1928–1931***

[1] Cairnes. *Geology and Mineral Deposits of Bridge River Mining Camp,* 77, and B.C. Department of Mines. *Index No. 3,* Queen's Printer, 1955, 193.
[2] *British Columbia Minister of Mines Reports,* 1909, K144, and 1910, K141–142, K149.
[3] Ibid., 1927, C215–216.
[4] B.C. Department of Mines. *Index No. 3,* Queen's Printer, 1955, 193.
[5] *The Lillooet Prospector,* September 22, 1916.
[6] Ibid., 1916, K268: 1917, F231, and 1918, K231.
[7] Ibid., 1925, A173.
[8] Public Archives of B.C., Registrar of Companies File 2984 (1897).
[9] PABC, Registrar of Companies GR 1526, Microfilm B5257, Lorne Gold Mines Limited, File 10064, 634 entries.
[10] Ibid., items 331, 332 and 340.
[11] Ibid., item 161.
[12] Stobie, Forlong's "Mining Review" pamphlet of December 1928.
[13] *Minister of Mines,* 1928, C216–218 and C432.
[14] Ibid., 1929, C231–233.
[15] *Canadian Mining Journal,* December 20, 1929, 1222.
[16] *The Financial Post,* November 14, 1929.
[17] *The Globe,* Toronto, January 31, 1930.
[18] Ibid., November 7–10, 1930.
[19] Ibid., March 19, 1931.
[20] *British Columbia Miner,* May 1930, 8–9.
[21] *Minister of Mines,* 1930, A202.

[22] PABC, Lorne Gold microfilm, item 262.
[23] *The Daily Province,* February 5, 1931.
[24] Ibid., March 25 and 27, 1931.

***Bralorne: Austin Taylor Takes Control, 1931–1932***

[1] PABC, Registrar of Companies, Corporate File 2329A.
[2] *The Financial Post,* April 14, 1934.
[3] Ada Ashby in *Bridge River-Lillooet News,* October 29, 1936.
[4] Joralemon, *Aventure Beacons,* 333.
[5] Cleveland interview.
[6] *Minister of Mines,* 1932, A222, and *The Miner,* November 1932, 335–336.
[7] *The Miner,* April 1937, 62–63.
[8] *The Daily Province,* July 20, 1932.
[9] Ibid., August 5, 1932.
[10] *The Miner,* June 1933, 537–538.
[11] *The Daily Province,* June 15, 1933.
[12] *Canadian Mining Journal,* June 1933, 246, and Joralemon, *Adventure Beacons,* 337.
[13] Clara Lambert in *Bridge River-Lillooet News,* February 20, 1948.
[14] Joralemon, "Veins and Faults in the Bralorne Mine," *Transactions of the American Institute of Mining and Metallurgical Engineers,* v. 115 (1935), 90–103.
[15] *Canadian Mines Handbook 1935,* 38–39.

***Working in the Mines***

[1] *Minister of Mines,* 1933, A263.
[2] "Toots" Chenoweth, *Western Miner,* December 1964, 64.
[3] *Bridge River-Lillooet News,* August 27, 1935.
[4] Hutchings album.
[5] *Minister of Mines* reports.
[6] Ibid., 1934, G43.
[7] Ibid., 1938, G44.
[8] Ibid., 1935, G38.

***Company Towns***

[1] Hutchings diary, March 8, 1936.
[2] Pioneer Chronology, 15.
[3] Bralorne Mines Annual Reports 5 (1935) and 7 (1937).
[4] *Bridge River Lillooet News,* March 25, 1937.

***Developments at the Mines***

[1] James and Emmons interviews.
[2] *Minister of Mines,* 1932, A223.
[3] Ibid., 1934, F28, and *The Northern Miner,* June 20, 1935.
[4] *The Vancouver Province,* July 31, 1935.
[5] *Minister of Mines,* 1935, F55.
[6] Ibid., 1933, A265.
[7] *Canadian Mines Handbook 1935,* 38. Bralco Limited, a private company incorporated in March 1933 took over the holdings of Bralco Development and

Investment Company, Limited which was wound up. Share interests were unchanged.

[8] *Bralorne Mines Limited,* Fourth Annual Report for the year ended 31 December, 1934.
[9] *The Vancouver Sun,* April 30, 1935.
[10] Ibid.
[11] Miller, Hedley and Poole interviews.
[12] *Bridge River-Lillooet News,* July 16, 1935.
[13] *Bralorne Mines,* 6th Annual Report for the year ended December 31, 1936.
[14] *Bralorne Mines,* 6th Annual Report for the year ended December 31, 1939.
[15] *Bralorne Mines,* 10th Annual Report for the Year Ending December 31, 1940.
[16] *Bridge River-Lillooet News,* August 8, 1941.
[17] *The Miner,* September 1941, 48.
[18] *Bridge River-Lillooet News,* September 5, 1941.
[19] *Minister of Mines,* 1941, A58.
[20] *Bralorne Mines,* 11th Annual Report for the Year Ending December 31, 1941.

***Stock Market Capers***

[1] Stead interview.
[2] Josephson, *The Money Lords,* 119.
[3] Ibid., 117.
[4] D. Stone, pers. comm. 1988.
[5] *The Financial Post,* November 3, 1934.
[6] Brooks. *Once in Golconda,* 79.
[7] Ibid., 122.
[8] *The New York Times,* May 12, 1961, 29.
[9] Josephson. *Infidel in the Temple,* 88.
[10] Ibid., *Minister of Mines,* 1932, A223.
[11] *The Daily Province,* January 15, 1933.
[12] Brooks. *Once in Golconda,* 194–196.
[13] *The Northern Miner,* April 25, 1935.
[14] *Minister of Mines,* 1932, A219.
[15] de Hullu. *Bridge River Gold,* 33.
[16] *The Financial Post,* June 9, 1934, 20.
[17] *Vancouver News Herald,* January 5, 1934.
[18] *The Vancouver Sun,* April 30, 1935.
[19] Cleveland interview and *The Financial Post,* September 24, 1932, 16.
[20] W.A. Hutchings diary.

***High-Grade and High-Grading***

[1] Eklof interview
[2] Pearson interview.
[3] Hedley interview.
[4] Ingram interview.
[5] *Minister of Mines,* 1895, 667.
[6] *Bridge River-Lillooet News,* March 16, 1939.
[7] *The Daily Province,* November 1 and 4, 1932.

[8] *Bridge River-Lillooet News,* June 4, 1936.
[9] Ibid., March 2, 1939.
[10] Ibid., March 16, 1939.
[11] Ibid., May 25, 1939.

***Strikes Won and Lost***

[1] Dunn interview.
[2] *Bridge River-Lillooet News,* May 16 and 23, 1935.
[3] *The Miner,* April 1937, 38.
[4] *Bridge River-Lillooet News,* April 6, 1939.
[5] Ibid., April 27, 1939.
[6] University of B.C. Archives, Cameron papers VF262a "Events Leading to Strike."
[7] *The Bridge River Miner,* November 11, 1939.
[8] Ibid., & *Bridge River-Lillooet News,* October 12, 1939.
[9] *Bridge River-Lillooet News,* October 12, 1939.
[10] Stanton manuscript prepared for *Never Say Die!*
[11] Ibid.
[12] *The Vancouver Sun,* November 16, 1939, and *Bridge River-Lillooet News,* November 16 and 23, 1939.
[13] *Bridge River-Lillooet News,* December 14, 1939.
[14] Stanton manuscript.
[15] de Hullu, *Bridge River Gold,* 84–85.
[16] *The Evening Province,* February 27 to March 1, 1940.
[17] *The Vancouver Sun,* March 6, 1940.
[18] *Bridge River-Lillooet News,* March 14, 1940.
[19] *The Vancouver Sun,* March 11, 1940.
[20] *Bridge River-Lillooet News,* March 14, 1940.
[21] Nomland statement, in Cameron Papers, VF 262A.

***Settlements and Characters: Pioneer to Haylmore***

[1] *The Vancouver Sun,* Bridge River Section, August 19, 1936, 1.
[2] *Bridge River-Lillooet News,* July 30, 1935, and June 1, 1939.
[3] de Hullu, *Bridge River Gold,* 56.
[4] Ibid., 83.
[5] *Bridge River-Lillooet News,* July 5, 1934.
[6] Ibid., January 30, 1936.
[7] Ibid., November 9, 1939.
[8] Ibid., November 1 and 8, 1946.
[9] PABC, Registrar of Companies, Lorne Amalgamated Mines Ltd., File 2984 (1897).
[10] Land Registry Kamloops, Consolidated Register of Indefeasible Fees, vol. 26, Noelton claim L5456 and Mining Recorder, Lillooet, Bill of Sale Book 5, p. 220: release dated August 30, 1933. Taylor had better luck with his Eagle group of four claims. The Eagle No. 1 and the Audrey Fraction went to Bralorne in November 1937 for 6,000 Bralorne shares, $40,000 and 25 percent of the gold produced, and the Eagle and Eagle Fraction were sold to Pioneer in June 1943 for $75,000. (Bralorne file, B.C. Registrar of Companies, Corporate Records, and Pioneer Chronological Table of Events.)

[11] "Memoirs of Angus Davis, XI," *Western Miner,* May 1950, 36–37.
[12] *Bridge River-Lillooet News,* November 22, 1934.
[13] Branca interview.
[14] *Bridge River-Lillooet News,* August 2 and October 11, 1934.
[15] Hutchings diary.
[16] *Bridge River-Lillooet News,* June 17, 1937.
[17] Branca interview.
[18] de Hullu, *Bridge River Gold,* 53.
[19] *Bridge River-Lillooet News,* December 6, 1934.
[20] Keddell, *The Newspapering Murrays,* 124.
[21] Hutchings diary.
[22] Ibid., and Frykberg interview.
[23] McCann, *Geology and Mineral Deposits,* 83.
[24] *Bridge River-Lillooet News,* October 10, 1935.
[25] Harris, *Halfway to the Goldfields,* 43.
[26] Hutchings album.
[27] Cairnes, *Geology and Mineral Deposits,* 48, and *Bridge River-Lillooet News,* April 4, 1935.
[28] *Bridge River-Lillooet News,* Special number, October 29, 1936.
[29] Hutchings album.
[30] *The Vancouver Sun,* Bridge River section, August 19, 1936.

*The Road to the Mines*
[1] *The Miner,* August 1934, 359–360.
[2] Railton interview.
[3] de Hullu, *Bridge River Gold,* 41ff.
[4] *Bridge River-Lillooet News,* June 14, 1934.
[5] Railton interview.
[6] *Minister of Mines,* 1933, A263.
[7] Railton interview.
[8] *Bridge River-Lillooet News,* August 31, 1939.
[9] Ibid., January–February 1935.

*Policing the Bridge River*
[1] *Bridge River-Lillooet News,* August 27, 1935, and *The Evening Province,* August 8, 1935.
[2] *Bridge River-Lillooet News,* November 7, 1935.
[3] *Bridge River-Lillooet News,* May 29, and June 12, 1942.

*Big Bill Davidson's Minto Mine and Townsite*
[1] Mrs. Jack Simpson, quoted in de Hullu, *Bridge River Gold,* 32.
[2] *Minister of Mines,* 1932, A217.
[3] Hutchings diary.
[4] *The Miner,* February 1935, 20, and *The Vancouver Sun,* November 9, 1934.
[5] *Canadian Mines Handbook 1935,* 167.
[6] *News Herald,* September 28, 1935.
[7] *The Vancouver Sun,* October 30, 1935.

[8] Ibid., February 6, 1936.
[9] Ibid., Bridge River section, August 19, 1936, 4.
[10] King interview.
[11] Hutchings diary and album.
[12] *The Vancouver Sun,* Bridge River Section, August 19, 1936, 11.
[13] Mrs. Jack Simpson quoted in de Hullu, *Bridge River Gold,* 32.
[14] *Bridge River-Lillooet News,* September 10, 1936.
[15] Ibid., September 17, 1936.
[16] Ibid.
[17] Ibid., July 8, 1937.
[18] *Minister of Mines,* 1936, F3–6.
[19] *The Vancouver Sun,* October 29, 1937.
[20] *Gold in British Columbia,* Preliminary Map No. 64, B.C. Ministry of Energy, Mines and Petroleum Resources, 1986.

*Summer Celebrations and Hockey Rivalries*

[1] *Bridge River-Lillooet News,* May 28, 1936.
[2] Ibid., June 1, 1939.
[3] *Ibid.,* July 5, 1934.
[4] Ibid., September 13, 1934.
[5] Hutchings album.
[6] Hutchings diary, January 26–27, 1936.
[7] Ibid., February 12, 1936.
[8] *Bridge River-Lillooet News,* February 27, 1936.
[9] Hutchings diary, February 18, 1936.
[10] *Bridge River-Lillooet News,* March 18, 1937.
[11] King interview.
[12] Quoted in *Bridge River-Lillooet News,* March 10, 1938.
[13] Ibid., December 1, 1938.
[14] Ibid., February 9, 1939.
[15] Ibid., February 16, 1939.
[16] Ibid., February 23, 1939.

*Wartime Changes, 1942–1945*

[1] *Bridge River-Lillooet News,* March 5, 1948.
[2] Ashmore interview.
[3] *Minister of Mines,* 1942, A55–56 and 1943, A60.
[4] Poole interview.
[5] Pioneer Gold Mines of B.C. Annual Reports.
[6] Branca interview.

*Gold Mining: A Troubled Industry, 1945–1958*

[1] Bralorne Mines Ltd., Letter to Shareholders of June 18, 1946.
[2] *Western Miner,* December 1946, 60.
[3] Pioneer Gold Mines of B.C. Annual Reports.
[4] Poole interview.
[5] Poole interview.

[6] *Bridge River-Lillooet News,* June 2, 1949.
[7] Ibid., August 18, 1949.
[8] Ibid., October 13, 1949 and Moore, *Angelo Branca,* 92.
[9] Poole interview.
[10] *Minister of Mines,* 1948, A97–102.
[11] Keir interview.
[12] *Minister of Mines,* 1954, A102–103, and 1955, 33.
[13] Ministry of Mines, *Geology, Exploration and Mining, 1969,* 187–188.
[14] *Bridge River-Lillooet News,* October 20, 1955.
[15] Ibid., September 29, 1955.
[16] Ibid., October 20, 1955.
[17] Ibid., November 14, 1957.

***Franc Joubin Takes Charge, 1958–1963***

[1] *Minister of Mines,* 1947, A134 and 227–230.
[2] Joubin, *Not for Gold Alone,* 44–157.
[3] Pioneer Gold Mines of B.C., Notice of Extraordinary General Meeting, February 25, 1959.
[4] *Western Miner and Oil Review,* May 1962, 36–43, and statement issued on the pouring of gold brick No. 3,000 on September 28, 1963.

***The Final Years, 1963–1971***

[1] *Minister of Mines,* 1965, 144.
[2] Ibid., 1966, 138–140.
[3] Ibid.
[4] *Western Miner,* June 1969, 20.
[5] *The Province,* September 1, 1971, 11–12.
[6] B.C. Ministry of Mines, *Geology, Exploration and Mining 1971,* 308.
[7] *Gold in British Columbia,* 1986.

***Afterword***

[1] Canadian military disability pensions, tax free and fully indexed, have increased almost 23-fold.
[2] *Bralorne Resources Limited,* Annual Reports for 1980 and 1987.
[3] Whiting interview.
[4] *Avino Mines & Resources Ltd.,* 1994 Annual Report.
[5] *Bralorne-Pioneer Gold Mines Ltd.,* 1995 Annual Report.
[6] Ibid.

# Bibliography


*Personal Communications*

A.H. "Al" Abbott, Philip Ashmore, Kathleen Bean, Mona Bohemier, Wilfred Bouvette, John Branca, Arnold Buhler, Charles Campbell, Jr., Courtney Cleveland, Irving Commons, Cleve Cunningham, Evelyn Cunningham, J.M. "Jack" Currie, Emma de Hullu, John DeLeen, Ruth Dubois, W. St.C. Dunn, R. "Bob" Eklof, E.F. "Ed" Emmons, Julie Frykberg, Norman Gladstone, Mathew S. Hedley, Frank Holland, Stuart Holland, Irene Howard, Tom Illidge, Jr., Don Ingram, Ruth James, Franc Joubin, Mike Jukich, Georgina Keddell, Jim and Helen Keir, Donald M. King, Egil Lorntzsen, Donald C. McKechnie, C.M. "Cy" and Mary Manning, Elaine Miller, G.P. Mitchell, Masse Mitchell, Beverly Orgnacco, Margaret Pasacreta, J.M. "Jack" Pearson, Allan W. Poole, Mike Purcello, S.S. "Sid" Railton, Mary Renshaw, Elsie Ross, David A. Sloan, John K. Sloan, Victor Spencer, Jr., John Stanton, Gordon Stead, David Stone, Frank Whiting, John Williams, Peter Wiseman, Louis Wolfin.

*Archives*

British Columbia Archives and Records Service, Victoria.
British Columbia Registrar of Companies, Corporate Records (on microfilm).
City of Vancouver Archives, Pioneer Gold Mine folder.
University of British Columbia.
    William J. Cameron papers.
    John Stanton papers.
    International Union of Mine, Mill and Smelter Workers records.
British Columbia & Yukon Chamber of Mines, clippings files.
Other Unpublished Material
    Hutchings, W.A. Diaries, 1934–1936 and photo album.
    James, D.H., and Weeks, J.P. *Bridge River Mineral Area,* September 1961, mimeographed, 14 p.
    Pioneer Gold Mines of B.C. Limited. *Chronological Table of Events Relating to Pioneer Mine.* [To March 31, 1954], mimeographed.

*Newspapers and Periodicals*

*Bridge River-Lillooet News,* Lillooet, March 1, 1934–
*Bridge River Miner,* September 22, 1939 to February 14, 1940 [6 numbers].
*The Financial Post,* Toronto.
*The Lillooet Prospector,* Lillooet, November 24, 1911 to April 27, 1917.
*The Miner,* began as *British Columbia Miner* and in later years *Western Miner.* 1927–1984, especially the Pioneer Mine number of August 1934 and the Bralorne number of April 1937.
*The Northern Miner,* Toronto.
*The Prospector,* Lillooet, July 14, 1898 to December 7, 1905 [incomplete].
Vancouver newspapers.

***Government Publications***

British Columbia Department of Mines [name varies]
*Annual Reports* 1874 to 1968.

*Geology, Exploration and Mining in British Columbia,* 1969–71.

Brewer, Wm. M. "Lillooet Mining Division." *B.C. Minister of Mines Annual Report,* 1913, p. K246–273.

Church, B.N. "Geology and Mineralization of the Bridge River Mining Camp." B.C. Ministry of Energy, Mines and Petroleum Resources, *Geological Fieldwork 1986,* Paper 1987-1, p. 23–33.

Leitch, C.H.B., and Godwin, C.I. "Geology of the Bralorne-Pioneer Gold Camp." B.C. Ministry of Energy, Mines and Petroleum Resources, *Geological Fieldwork 1985,* Paper 1986-1, p. 311–316.

—. "The Bralorne Gold Vein Deposit: An Update." *Geological Fieldwork 1986,* Paper 1987-1, p. 35–38.

Ministry of Energy, Mines and Petroleum Resources: *Gold in British Columbia.* Preliminary Map No. 64, 1986.

Robertson, William Fleet. "Lillooet Mining Division." *B.C. Minister of Mines Annual Report* 1910, p. K134–148.

***Geological Survey of Canada, Ottawa***

Cairnes, C.E. *Geology and Mineral Deposits of Bridge River Mining Camp, British Columbia.* Geological Survey of Canada, Memoir 213, 1937.

McCann, W.S. *Geology and Mineral Deposits of the Bridge River Map-area, British Columbia.* Geological Survey of Canada, Memoir 180, 1922.

***Books, articles, and pamphlets***

Avino Mines & Resources Ltd. *1994 Annual Report.*

Bellamy, J., and Saleken, L.W. "Bralorne Gold Mine." *Some Gold Deposits in the Western Canadian Cordillera.* Field Trip Guidebook 4, Victoria: Geological Association of Canada, 1983, p. 23–39.

Bosustow, R. "Development and Present Operation of the Bralorne Mines." *The Miner,* November 1933, p. 711–713.

—. "A history of the Bralorne Mine." *The Miner,* April 1937, p. 36–37.

Bralorne Mines Limited. *Annual Reports,* 1931–

Brooks, John. *Once in Golconda: A True Drama of Wall Street 1920–1938.* New York: Harper and Row, 1969.

Cain, Harry J. and Schutz, Paul. "Milling at Pioneer Gold Mines." *The Miner,* November 1932, p. 331–336.

Chenoweth, E.J. and Hill, Henry H. "Bralorne Mining Practice." *The Miner,* April 1937, p. 47–54.

Cirkel, Fritz. "The Bridge River Gold Mining Camp." *Journal of the Canadian Mining Institute,* 1900, p. 21–29.

Cleveland, Courtney. "Geology of the Bralorne." *The Miner,* April 1937, p. 39–42.

Cleveland, Courtney E., and the Pioneer Staff. "Geology of the Bralorne and Pioneer Mines." *Transactions of the Canadian Institute of Mining and Metallurgy,* v. XLI, 1938, p. 12–27.

Craig, Andy A. *Trucking: British Columbia's Trucking History.* Saanichton: Hancock House, 1977.

de Hullu, Emma. *Bridge River Gold.* Bralorne: Bralorne Pioneer Community Club, 1967.

Edgeworth, W.A., and Foerster, F. "Accident Prevention and Dust Control at Bralorne." *The Miner,* May 1940, p. 32–36.

Edwards, Irene. *Short Portage to Lillooet.* Lillooet: the author, 1978.

Gibson, Swanston, and Poole, Allan. "Bralorne – Its History and Geology." *Western Miner,* December 1945, p. 40–44.

Gray, Fred E. "Bralorne Mill of Bralorne Mines Limited." *The Miner,* November 1932, p. 335–36.

Gray, Fred E., and Almstrom, A.A. "Milling at the Bralorne." *The Miner,* April 1937, p. 54–57.

Harris, Lorraine. *Halfway to the Goldfields.* North Vancouver: J.J. Douglas, 1977.

Hedley, M.S. "Geologic Structure at Bralorne Mine." *The Miner,* April 1935, p. 22–25.

James, Howard T. "Features of Pioneer Geology." and "Mining Methods at the Pioneer." *The Miner,* Pioneer number, August 1934, p. 342–50.

Joralemon, Ira B. "Veins and Faults in the Bralorne Mine." *Transactions of the American Institute of Mining and Metallurgical Engineers,* v. 115, 1935, p. 90–103.

—. *Adventure Beacons.* New York: Mining and Metallurgical Society of America, 1976, 487 p.

Josephson, Matthew. *Infidel in the Temple.* New York: Knopf, 1967.

—. *The Money Lords.* New York: Weybright and Talley, 1972.

Joubin, Franc R. "Bralorne and Pioneer Mines." in *Structural Geology of Canadian Ore Deposits.* Montreal: Canadian Institute of Mining and Metallurgy, 1948, p. 168–177.

—. "Bridge River Pioneer." *Western Miner and Oil Review,* August 1958, p. 36–40.

Joubin, Franc R., and Smyth, D. McCormack. *Not for Gold Alone.* Toronto: Deljay Publications, 1986.

Keddell, Georgina. *The Newspapering Murrays.* Lillooet: Lillooet Publishers, 1974.

Lambert, Clara. "Dig, Brother, Dig!" *Bridge River-Lillooet News,* February 13, 20 and 27, and March 5, 1948.

Leitch, C.B.H., van der Heyden, P., Godwin, C.I., Armstrong, R.L., and Harakal, J.E. "Geochronometry of the Bridge River Camp, southwestern British Columbia." *Canadian Journal of Earth Sciences,* v. 28, 1991, p. 195–208.

Maiden, Cecil. *Lighted Journey: The Story of the B.C. Electric.* [Vancouver, c1948].

Moore, Vincent. *Angelo Branca, "Gladiator of the Courts."* Vancouver, Douglas and McIntyre, 1981.

Pioneer Gold Mines of B.C. Limited. *Annual Reports,* 1929 to 1958.

Poole, Allan W. "The Geology and Analysis of Vein and Fault Structure of the Bralorne Mine." *Canadian Institute of Mining and Metallurgy Bulletin,* November 1955, p. 733–737.

Richards, R.H., and Locke, C.E. *A Text Book of Ore Dressing.* New York: McGraw-Hill, 1925.

Schutz, Paul and Spry, Russell J. "The Pioneer Mill." *The Miner,* Pioneer Mine number, August 1934, p. 350–353.

Sloan, David. "History and Early Development of the Pioneer Gold Mine." *The Miner,* Pioneer number, August 1934, p. 339–342.
Stanton, John. *Never Say Die! The Life and Times of a Pioneer Labour Lawyer.* Ottawa: Steel Rail, 1987.
Stephens, Fred H. "A New Mill for Bralorne." *Western Miner and Oil Review,* May 1962, p. 36–43.
Thrush, Paul W., compiler and editor. *A Dictionary of Mining, Mineral and Related Terms.* Washington: U.S. Department of the Interior, 1968.
Wood, George Alan. *The Bridge River Region – A Geographical Study.* M.A. thesis, University of British Columbia, 1949.
Woollacott, Arthur P. "Gold in Lillooet." *Maclean's Magazine,* March 1, 1934, p. 13 and 31.

# Index

*Page numbers in italics refer to illustrations.*

Lewis Green is a retired geologist who worked with the Geological Survey of Canada and later in the mineral exploration industry. His previous books are *The Gold Hustlers* (1977) and *The Boundary Hunters* (1982).